# Case Workbook

to accompany

# Human Genetics
## *Concepts and Applications*

### Sixth Edition

**Ricki Lewis**

Boston   Burr Ridge, IL   Dubuque, IA   Madison, WI   New York   San Francisco   St. Louis
Bangkok   Bogotá   Caracas   Kuala Lumpur   Lisbon   London   Madrid   Mexico City
Milan   Montreal   New Delhi   Santiago   Seoul   Singapore   Sydney   Taipei   Toronto

**The McGraw·Hill Companies**

Case Workbook to accompany
HUMAN GENETICS: CONCEPTS AND APPLICATIONS, SIXTH EDITION
RICKI LEWIS

Published by McGraw-Hill Higher Education, an imprint of The McGraw-Hill Companies, Inc., 1221 Avenue of the Americas, New York, NY 10020. Copyright © 2005, 2003, 2001, 1994 by The McGraw-Hill Companies, Inc. All rights reserved.

No part of this publication may be reproduced or distributed in any form or by any means, or stored in a database or retrieval system, without the prior written consent of The McGraw-Hill Companies, Inc., including, but not limited to, network or other electronic storage or transmission, or broadcast for distance learning.

This book is printed on acid-free paper.

1 2 3 4 5 6 7 8 9 0 DCD DCD 0 9 8 7 6 5 4

ISBN 0-07-284606-2

www.mhhe.com

# TABLE OF CONTENTS

**PART ONE**      **INTRODUCTION**

**CHAPTER 1**      OVERVIEW OF GENETICS
Genetics in the News, **2**

**CHAPTER 2**      CELLS
Carnitine-acylcarnitine translocase deficiency, **3**
Combined factors V and VIII deficiency, **5**

**CHAPTER 3**      DEVELOPMENT
Embryos and fetuses in research, **7**

**PART TWO**      **TRANSMISSION GENETICS**

**CHAPTER 4**      MENDELIAN INHERITANCE
Acrocephalosyndactyly, **10**
Carnosinemia, **12**
Huntington-like disorder, **14**
Restless leg syndrome, **15**
Schneckenbecken dysplasia, **16**

**CHAPTER 5**      EXTENSIONS AND EXCEPTIONS TO MENDEL'S LAWS
Enamel hypoplasia, **17**
Epidermolysis bullosa, **19**
Hair and eye color, **21**
Thrombocytopenia and absent radius syndrome, **23**

**CHAPTER 6**      MATTERS OF SEX
Anhidrotic ectodermal dysplasia, **24**
Blue diaper syndrome, **25**
Chronic granulomatous disease, **26**
Congenital muscular dystrophies, **28**
Intersex, **30**

**CHAPTER 7**      MULTIFACTORIAL TRAITS
Cleft lip with or without cleft palate, **31**
Complex traits among the Hutterites, **33**

**CHAPTER 8**      THE GENETICS OF BEHAVIOR
Alcoholism, **35**

**PART THREE**      **DNA AND CHROMOSOMES**

**CHAPTER 9**      DNA STRUCTURE AND REPLICATION
DNA replication, **38**

| | |
|---|---|
| CHAPTER 10 | GENE ACTION AND EXPRESSION<br>Alpha$_1$-antitrypsin deficiency, **40**<br>Hypoxia-inducible factor I, **42** |
| CHAPTER 11 | GENE MUTATION<br>Bloom syndrome, **44**<br>DNA repair, **46**<br>Gyrate atrophy, **47**<br>Open-angle glaucoma, **48**<br>Otospondylomegaepiphyseal dysplasia (OSMED), **49**<br>Tay-Sachs disease, **50**<br>Von Willebrand disease, **52** |
| CHAPTER 12 | CHROMOSOMES<br>DiGeorge syndrome, **54**<br>Down syndrome, **56**<br>Tetrasomy 12, **57**<br>Turner syndrome, **58**<br>Williams syndrome, **59** |
| **PART FOUR** | **POPULATION GENETICS** |
| CHAPTER 13 | WHEN ALLELE FREQUENCIES STAY CONSTANT<br>The ice maiden, **62** |
| CHAPTER 14 | CHANGING ALLELE FREQUENCIES<br>Type III 3-methylglutaconic aciduria, **63**<br>Ulnar-mammary syndrome, **65** |
| CHAPTER 15 | HUMAN ORIGINS AND EVOLUTION<br>Novelty seeking and ADHD, **66** |
| **PART FIVE** | **IMMUNITY AND CANCER** |
| CHAPTER 16 | GENETICS OF IMMUNITY<br>Five little piggies, **69** |
| CHAPTER 17 | GENETICS OF CANCER<br>Acute T cell leukemia, **71**<br>Li-Fraumeni family cancer syndrome, **72**<br>Multiple endocrine neoplasia, **74**<br>Thyroid cancer, **76** |
| **PART SIX** | **GENETIC TECHNOLOGY** |
| CHAPTER 18 | GENE MODIFICATION<br>Hemophilia A and B, **78**<br>Infertility drugs, **80** |

| | | |
|---|---|---|
| CHAPTER 19 | GENE THERAPY AND GENETIC COUNSELING | |

Gene doping, **81**
Hemophilia B, **82**
Newborn screening, **84**

| | | |
|---|---|---|
| CHAPTER 20 | AGRICULTURAL BIOTECHNOLOGY | |

Designer potatoes, **86**
Transgenic tobacco, **87**

| | | |
|---|---|---|
| CHAPTER 21 | REPRODUCTIVE TECHNOLOGIES | |

Charcot-Marie-Tooth disease, **89**
Male infertility, **90**

| | | |
|---|---|---|
| CHAPTER 22 | THE HUMAN GENOME PROJECT AND GENOMICS | |

Diffuse large B cell lymphoma, **91**
Muscle cell DNA microarray, **93**

**PART SEVEN**     **CONNECTIONS AND SYNTHESIS**

Argininemia, **96**
BPES syndrome, **98**
Complement component 2 deficiency, **100**
Dilated cardiomyopathy, **102**
Ehlers-Danlos syndrome type IV, **104**
Familial mental retardation (ATR-16), **106**
Hereditary multiple exostoses, **108**
Lysinuric protein intolerance, **110**
Muscular dystrophy, **111**
Nephrolithiasis, **113**
Prenatal DNA microarray screen, **115**
Pseudohermaphroditism, **117**
Silver-Russell syndrome, **119**
Smith-Lemli-Opitz syndrome, **120**
Stem cells in the heart, **122**
Tangier disease, **125**
Townes-Brocks syndrome, **127**
Venous thrombosis, **129**

Appendix A The Genetic Code, **131**
Appendix B Pedigree Symbols, **133**

# PREFACE

The study of human genetics is the study of inherited variations among people. Often these variations cause diseases. The goal of this Case Workbook is to place the information in *Human Genetics: Concepts and Applications* into a real context. Most of the cases come from the medical or scientific literature, with an emphasis on the unusual. Many introduce information and details that are not found in textbooks. Each case includes information, the source in either Online Mendelian Inheritance in Man and/or journals, and questions. The cases are organized to follow the chapter outline of the textbook, with a final part devoted to "Connections and Synthesis." These final cases depart from the rather artificial division of topics in a textbook. They are an excellent means to study for a final exam.

Most of the questions require thinking and problem-solving, rather than straight recall of facts or terminology. This approach reflects the skills required of a genetic counselor – powerful observation, data interpretation, thinking and communicating. For years, the primary tools of the genetic counselor and medical geneticist were the pedigree and karyotype. Today, analysis of an inherited condition often embraces molecular information too. Applying Mendel's laws to predict recurrence risks and to identify carriers, as well as interpreting DNA microarray tests, requires applying the rules of probability, as well as logic. Would a person with a particular condition live long enough or feel well enough to have children? Under what conditions would an extremely rare disorder be unusually prevalent? Where in the body must a gene therapy be targeted to alleviate symptoms? More generally, which facts in a case are relevant to answering particular questions? Solving genetic problems may help you to get the most out of other types of information too. Have fun!

Thanks to Daniel Plyler of the University of North Carolina at Wilmington and Ruth Sporer of the University of Pennsylvania for their comments during the development of this edition of the Case Workbook.

Ricki Lewis

# PART ONE

# INTRODUCTION

**CHAPTER 1**  **OVERVIEW OF GENETICS**
  Genetics in the news

**CHAPTER 2**  **CELLS**
  Carnitine-acylcarnitine translocase deficiency
  Combined factors V and VIII deficiency

**CHAPTER 3**  **DEVELOPMENT**
  Embryos and fetuses in research

# GENETICS IN THE NEWS — CHAPTER 1

Chapter 1 is an overview of many of the topics in the textbook. Consult recent newspapers, magazines, or online sources and describe examples of the following phenomena:

A DNA chip

A prenatal test

A cancer diagnosis or treatment based on genetics

A medical condition for which a person has an increased risk because of inheriting a particular gene variant

A genetically modified plant or animal

A drug made possible using biotechnology

DNA fingerprinting in a forensics application

Genetic determinism

Controversy over the use of stem cells or cloning a human

Gene therapy

# CARNITINE-ACYLCARNITINE TRANSLOCASE DEFICIENCY     CHAPTER 2

Jim D. died at 4 days of age, 2 days after suffering cardiac arrest. The pregnancy had been uneventful, and he had seemed normal at birth. His older sister was healthy. Two years after Jim's death, his parents had another son.

Like Jim, Kerry seemed healthy at birth. But at 36 hours, his heart rate fell, he had a seizure, and he stopped breathing. Doctors were able to resuscitate him. Knowing Jim's history, the doctor ordered several tests and discovered excess long-chain fatty acids in Kerry's blood. These are components of triglyceride fats that are broken down by mitochondria in the liver, to provide glucose for energy during starvation. Nurses had noted that Kerry had not eaten since birth.

Kerry was fed a special infant formula low in fatty acids, and his parents were instructed to feed him often. Despite this treatment, the child was frequently hospitalized for vomiting, lethargy, an enlarged liver, and poor muscle control. Once he went into a coma after a feeding was delayed. Each time he became ill, it would take two to three weeks for the boy to recover. His weakness and fatigue worsened, and triglycerides began to accumulate in his muscles. His liver enlarged. Finally, just before his third birthday, Kerry died in the hospital of respiratory failure.

The brothers had a rare inborn error of metabolism called carnitine-acylcarnitine translocase deficiency. Their cells lack an enzyme produced in mitochondria.

**SOURCE:**  OMIM 212138

**WORKSHEET:**

1. Why is this disorder considered to be lethal in an evolutionary sense, even though it is not fatal for a few years to the individual?




2. A symptom that is consistent with the site of abnormality in the cell is

_____.

3. The special food that Kerry received attempted to lessen the amount of the fatty acids that built up in his blood. Another type of approach to correcting the underlying biochemical defect might be to

_____

_____.


4. The children's mother is taking a course in biology, and she learns that mitochondria are passed from the mother only. When the doctor explained that her sons had abnormal mitochondria, she began to feel guilty for having transmitted the disorder to them. What is the evidence that she is incorrect in thinking that her mitochondrial gene made the boys ill?

# COMBINED FACTORS V AND VIII DEFICIENCY             CHAPTER 2

Blood clotting requires 11 enzyme-catalyzed chemical reactions. Several dozen families have poor clotting because of inherited deficiencies of two clotting factors, V and VIII. Deficiency of factor V causes an autosomal recessive condition called parahemophilia, and factor VIII deficiency causes the more familiar X-linked hemophilia A. However, the combined deficiency is inherited as a distinct autosomal recessive condition. That is, it affects both sexes and is inherited from two carrier parents.

Researchers expected to find an inborn error of metabolism that affects an enzyme that functions in the clotting pathway before factors V and VIII act. Instead, they discovered a gene that encodes a carrier protein that shuttles certain glycoproteins between the ER and the Golgi apparatus for secretion. The protein normally transports factors V and VIII, which are similar glycoproteins.

**SOURCES:** OMIM 227310

Nichols, William C. and David Ginsburg. June 1999. From the ER to the Golgi: insights from the study of combined factors V and VIII deficiency. *The American Journal of Human Genetics* 64:1493.

**WORKSHEET:**

1. Clotting factors, enzymes, and carrier proteins have very different functions. How can they all be products of gene activity?




2. How would the original hypothesis of an earlier defect in a shared biochemical pathway explain the double clotting factor deficiency?

3. Describe what happens to the product of the gene that causes the double clotting factor disorder in the ER and the Golgi apparatus.

4. What is the evidence that a mutation in a single gene causes this double deficiency? That is, how does the inheritance pattern differ from what it would be if families inherited each condition independently?

# EMBRYOS AND FETUSES IN RESEARCH            CHAPTER 3

A human organism cannot survive outside the uterus, using current technology, prior to 24 weeks of gestation. The period of the embryo lasts until the end of the eighth week, by which time the rudiments of all organs have appeared, and the period of the fetus then extends until the time of birth. The status of the prenatal human is the topic of diverse arguments and unending debate, with opinions differing even within groups who are united in other matters.

Until the late 1990s, concern about using embryos in research centered around the problem of what to do with frozen "extras" *from in vitro* fertilization (see Chapter 21). Then in 1998, when two groups of researchers succeeded in culturing human embryonic stem (ES) cells, the possibility of using these cells in regenerative medicine and other applications created excitement in the scientific and medical communities, at the same time that it alarmed many in the general population. Vigorous debate ensued, continues, and likely will not cease, no matter what type of legislation passes. Below are some varied viewpoints on the status of embryos, which has bearings on the ethics of experimentation with embryos. Of course, not all people who follow a particular religion agree with the majority viewpoint on this highly controversial matter.

- Fifteenth century Islamic scholar Hajar al-Asqalani stated that the human fetus is "like a plant" and that for the first 40 days of prenatal development, the embryo is "as a drop of matter." The soul enters after 120 days.

- According to Jewish law, the embryo before 40 days has no moral status, and is "as if they were simply water."

- Buddhists and Hindus maintain that because a "transmigration of consciousness" occurs at conception, the embryo is a being.

- In the United Kingdom, Canada, Sweden and Finland, research can be conducted on embryos before 14 days of gestation. This is the time when the primitive streak establishes the body axis.

- In 1994, an advisory panel to the National Institutes of Health concluded that day 22, when the heart begins to beat, is "... the first time the embryo can be perceived (through ultrasound) by the outside world. ... despite experience with brain death, it is the beating heart that is most strongly perceived to be the difference between life and death."

- Biology textbooks avoid certain terms, such as referring to a pregnant woman as a mother, a fetus as an unborn child, or prenatal life. On the other hand, although an embryo could not survive outside of a uterus, it does carry out processes that are part of life, such as DNA replication and metabolism.

**SOURCE**: *The New York Times*, Science Times, December 18, 2001, has several articles from which this case was developed.

**WORKSHEET:**

1. Compare the structures that are present in the human embryo at the inner cell mass stage, day 14, and day 22.

2. Which characteristics or qualities do you think are important to consider in deciding whether or not an embryo or fetus should be afforded the same rights and considerations as a person?

3. Therapeutic cloning is a theoretical and therefore experimental procedure that uses cells from the inner cell mass to generate human embryonic stem cells. The ES cells are then used to culture tissues that can be used as grafts to help a person with a disorder such as Parkinson disease or a spinal cord injury. The rest of the embryo is then discarded. Do you think that this type of work should be legal? Cite a reason for your answer.

4. Do you think that early human embryos should be used in research to learn more about development? Cite an alternative to this approach, and its advantages and disadvantages compared to using human cells.

# PART TWO

# TRANSMISSION GENETICS

**CHAPTER 4 MENDELIAN INHERITANCE**
    Acrocephalosyndactyly
    Carnosinemia
    Huntington-like disorder
    Restless leg syndrome
    Schneckenbecken dysplasia

**CHAPTER 5 EXTENSIONS AND EXCEPTIONS TO MENDEL'S LAWS**
    Enamel hypoplasia
    Epidermolysis bullosa
    Hair and eye color
    Thrombocytopenia and absent radius syndrome

**CHAPTER 6 MATTERS OF SEX**
    Anhidrotic ectodermal dysplasia
    Blue diaper syndrome
    Chronic granulomatous disease
    Congenital muscular dystrophies
    Intersex

**CHAPTER 7 MULTIFACTORIAL TRAITS**
    Cleft lip with or without cleft palate
    Complex traits among the Hutterites

**CHAPTER 8 THE GENETICS OF BEHAVIOR**
    Alcoholism

# ACROCEPHALOSYNDACTYLY — CHAPTER 4

Wayne and Marge had always thought that their identical twin sons Wally and Todd had unusually large toes. This was cute when they were infants, but as they began to walk, their feet did not fit easily into shoes. The problem worsened as they grew older, but then they discovered a certain brand of running shoe that fit well. Wayne's mother Cecile recalled that she had similar problems fitting Wayne and his sister Colleen with shoes when they were children. Cecile's deceased husband George had very bizarre feet. Colleen and her husband Jack have a 4-year-old daughter, Leah, who has the family's large toes too.

Colleen is a geneticist. She suspected that Willy, George, Todd, Wayne, Leah, and herself have a form of acrocephalosyndactyly, an inherited disorder. She suggested that the affected relatives have their toes x-rayed and, as she suspected, they all had the unusual sign of this condition – double bones in each toe.

**SOURCE:** OMIM 105200

**WORKSHEET:**

1. This disorder is very rare. Of known affected families, there are no instances of consanguinity. Does this support a mode of inheritance of autosomal recessive or autosomal dominant inheritance? Why?

2. If the trait is inherited as an autosomal dominant, and Leah marries John, a ballet dancer with exquisite, slim toes, the probability that a child of theirs would inherit normal toes is_____.

3. Complete the pedigree of this family if the big toe trait is inherited as an autosomal dominant condition.

4. Complete the pedigree of this family if the big toe trait is inherited as an autosomal recessive condition, including individuals who must be carriers.

# CARNOSINEMIA — CHAPTER 4

Simon and Adrienne Jackson are healthy people who lived in a rural area with poor health services when they had their first child Benjie. He had seizures in infancy, and as he grew into toddlerhood, it became apparent that he was severely mentally retarded. He died at 26 months of age. Because he had never been diagnosed with a specific disorder, an autopsy was performed. His brain showed signs of great derangement, with nerve cells degenerating and missing. No diagnosis was made, and because no other relatives had been affected, a genetic problem was not suspected. The family physician assured the couple that the condition was not likely to repeat.

The Jacksons waited a few years, then had another child after moving to Chicago. Sadly, little Julie had the same symptoms as her brother. This time, the parents took her to a major medical center, where urine and cerebrospinal fluid tests revealed large amounts of a chemical called carnosine that consists of two types of amino acids, alanine and histidine, joined together. Digestion should have broken the carnosine down into the individual amino acids, which are small enough to enter the bloodstream. When a medical geneticist learned of Julie's test results, she tested the urine of the parents. Each had half the normal activity for an enzyme called carnosinase. Julie has and Benjie had an inherited disorder, carnosinemia.

**SOURCE:** OMIM 212200

**WORKSHEET:**

1. The mode of inheritance for carnosinemia in this family is

   _____.


2. What is the biochemical evidence that indicates the mode of inheritance?

3. The probability that Simon and Adrienne can conceive a child who does not inherit carnosinemia is _____.

4. The probability that Simon and Adrienne can conceive a child who is a carrier like they are is _____.

5. How might you devise a treatment for carnosinemia?




6. In one experiment on two children with carnosinemia, all sources of dietary protein with an alanine next to a histidine were eliminated from the diet. The children still excreted carnosine in the urine. What is an explanation for this finding?

# HUNTINGTON-LIKE DISORDER

**CHAPTER 4**

A newly discovered disease produces symptoms very much like those of Huntington disease (HD) -- loss of balance, uncontrollable movements, but also with intellectual impairment and seizures. The new disease is autosomal recessive and caused by mutation in a gene that maps to chromosome 4 in the same general region as the gene that causes HD, but does not have trinucleotide repeats. Age of onset is between one and four years, and it is fatal in childhood. The following pedigree was published in the journal article describing the disorder:

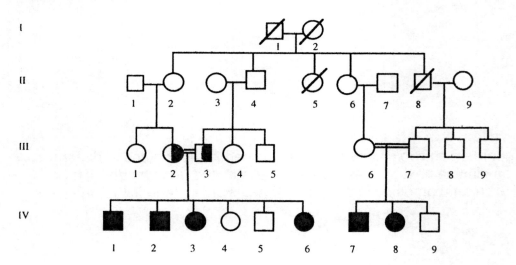

**SOURCE:** Kambouris, M. et al, February 2000. Localization of the gene for a novel autosomal recessive neurodegenerative Huntington-like disorder to 4p15.3. *The American Journal of Human Genetics* 66:445.

## WORKSHEET:

1. Which individuals in the pedigree are deceased?

2. If this condition is extremely rare, why did so many people in the fourth generation of this family inherit it?

3. Who must be carriers?

# RESTLESS LEG SYNDROME                                      CHAPTER 4

One in 20 people has restless leg syndrome, in which the legs feel tingly or aching nearly all the time, causing great daytime fatigue from interrupted sleep. A gene for the condition has been tentatively localized to chromosome 12. Another fairly common inherited condition causes extremely red hair and an inability to tan. The person has an unusual variant of the pigment molecule melanin, called "red" melanin. Both restless leg syndrome and red melanin are recessive, and their genes are on different chromosomes.

Suzanne and Michael met at the original Woodstock rock festival in the summer of 1969, drawn to each other because of their striking red hair and very pale skin. Neither has a family history of restless leg syndrome. A few years later they married, and in 1975 had Marvin and Gary, monozygotic (identical) red headed twins who do not tan.

Jackie and David were at Woodstock too. They each have dark brown hair, and David suffers from restless leg syndrome. Their daughter, Eileen, has dark brown hair and legs that aren't restless.

Gary and Eileen get together, and have a son, Todd. He has red hair, doesn't tan, and has restless leg syndrome.

## SOURCE:
Desautels, Alex et al. December 2001. Identification of a major susceptibility locus for restless legs syndrome on chromosome 12q. *The American Journal of Human Genetics* 69:1266

## WORKSHEET:

1. Which individuals must be heterozygous for restless leg syndrome?

2. Which individuals might be heterozygous for restless leg syndrome?

3, Which individuals must be heterozygous for the red melanin gene?

4. If Marvin marries a woman who has restless leg syndrome and red hair and cannot tan, what is the probability that a child of theirs will inherit both traits?

# SCHNECKENBECKEN DYSPLASIA CHAPTER 4

Stuart and Gloria have had four healthy children, five stillborn dwarves, and three miscarriages. In their most recent pregnancy, ultrasound showed dwarfism in the third month. The phenotype includes very short leg bones, small vertebrae, and a twisted pelvis. In German, "schneckenbecken" means "snail pelvis." The pedigree is shown below.

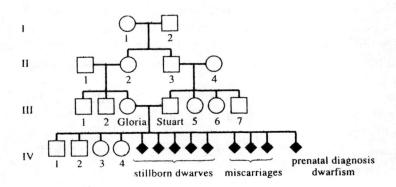

**SOURCE:** OMIM 269250

**WORKSHEET:**

1. Stuart and Gloria have 1/_____ of their genes in common.

2. The chance that individual IV2 is not a carrier is _____.

3. Which individuals could be carriers of this condition?

# ENAMEL HYPOPLASIA    CHAPTER 5

In enamel hypoplasia, holes and cracks appear around the crowns of the baby teeth. It is inherited as an autosomal dominant trait, but with incomplete penetrance and variable expressivity. Below is a pedigree showing the Barker family, which has this trait, and the Needlemeyer family, with whom they marry.

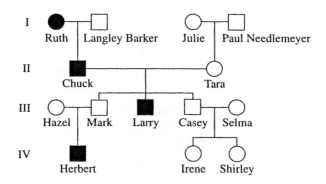

**SOURCE:** OMIM 600907

**WORKSHEET:**

1. From the pedigree, it appears that the individual who displays incomplete penetrance for this trait is _____.

2. How might the variable expressivity appear among the affected members of the Barker family?

3. Irene, as a baby, liked to go to sleep with a bottle of juice. The sugar washing over her teeth caused such terrible decay that the distinctive marks of enamel hypoplasia could not be seen. Explain how determining haplotypes for certain family members could reveal whether or not Irene inherited enamel hypoplasia.

4. The phenomenon of an environmental health problem, such as tooth decay caused by going to sleep with a juice bottle in the mouth, resembling a known inherited disorder is called a _____.

5. The probability that Herbert's child will inherit enamel hypoplasia is

_____.

6. The Barkers clean their attic and find old baby photos that shed some light on the family tooth anomaly. Ruth's mother Lucy had a toddler's grin full of odd shaped teeth. Her father Jerry had normal teeth. Ruth also discovers that her older brother Fred and older sister Anna had the abnormal teeth, but she remembers that her younger sister Lulubelle had beautiful, healthy teeth. Add this information to the pedigree.

# EPIDERMOLYSIS BULLOSA                                      CHAPTER 5

In epidermolysis bullosa (EB), the skin blisters very easily (see figure 2.11). Parents whose children have EB are chatting while waiting to see their dermatologist. The Renfrews have a son and daughter, Bryan and Cheri, who each have the dystrophic form of EB. This is caused by a recessive mutation in a gene encoding type VII collagen, a connective tissue protein that forms fibers that anchor the lower skin layer, the dermis, to the upper layer, the epidermis. It is a very disfiguring form of the illness. Bryan and Cheri, who are twins, have many blisters on their skin, which leave scars. Rita Renfrew is pregnant, and she and her husband Ronald are concerned that their next child will be affected too.

The Blackwells also have two children who have EB, Chad and Jeremy, aged 5 and 7. They have a milder, simplex form of the illness, with blisters that form on their hands and feet during warm weather but do not leave scars. Their mother Beulah also has the condition, which is inherited as an autosomal dominant trait. EB simplex is caused by a mutation in a gene on chromosome 5 that encodes keratin, a protein that is abundant in the epidermis.

Another couple without children sits in a corner. The Starkey's son Richard died in infancy of the severe basement membrane form of EB. He had inherited a recessive allele from each parent that made him unable to produce a protein called epiligrin that anchors the epidermis to a tissue layer called the basement membrane.

**SOURCES:** OMIM 131950, 226500, 226650, 226700

**WORKSHEET:**

1. Explain the genetic heterogeneity of EB. That is, how can similar phenotypes have different genotypes?




2. If Cheri Renfrew marries Jeremy Blackwell, the probability that their child inherits EB simplex is _____.

3. Draw a pedigree for the Renfrew family.



















4. Gene therapy may be possible for Chad and Jeremy, because keratinocytes (skin cells filled with keratin) can grow in culture, accept foreign genes, and be grafted onto a person. Why wouldn't this approach help Bryan and Cheri or future children of the Starkeys who inherit EB?

# HAIR AND EYE COLOR — CHAPTER 5

In Scandinavia, people who do not have blond hair and blue eyes stand out. A study considered hair and eye color in 100 families in Copenhagen, Denmark. In each of 50 couples (the F1 generation), one partner has the common blond hair and blue eyes, and the other has brown hair and brown eyes. The brown haired and eyed parents each have a parent who looks or looked like them, and the other parent has or had blond hair and blue eyes (the P1 generation). The fifty F1 parents produce 260 children (the F2 generation). The 260 children have the following phenotypes and frequencies:

| Phenotype | # of children |
|---|---|
| Brown hair, blue eyes | 4 |
| Blond hair, brown eyes | 6 |
| Brown hair, brown eyes | 120 |
| Blond hair, blue eyes | 130 |

Brown hair (H) is dominant to blond hair (h), and brown eyes (E) are dominant to blue eyes (b).

**SOURCES:** OMIM 227220, 601800

**WORKSHEET:**

1. What is the genotype or genotypes of the blond haired blue-eyed parents of the F1 generation?

2. Are the genes for these two traits linked or unlinked? How can you tell?

3. Explain or diagram how the unusual appearing children in the F2 generation arose.

4. Marker studies showed a LOD score for these two genes of 5.4. This means that the genes are

       a. located far apart on the same chromosome
       b. located very close together on the same chromosome
       c. located on either side of the centromere
       d. each present in three allelic variants
       e. located on different autosomes

# THROMBOCYTOPENIA AND ABSENT RADIUS SYNDROME     CHAPTER 5

Marjorie and Joe Winthrop are healthy, but only two of their six children are well. Michael, the eldest, was born without arms, and has an abnormal heart. Jacob, born next, is healthy. Marianne, the third eldest, suffered near-constant stomach cramps until her parents realized that she was allergic to cow's milk. She also has impaired blood clotting and excess eosinophils, a type of white blood cell. The next child, Adam, is healthy. The youngest two children, Mira and Pete, are twins. Pete is allergic to cow's milk and has a minor heart abnormality. When Pete's cardiologist asks for a family history, and Joe mentions the various symptoms of the children, the doctor realizes that the family has thrombocytopenia and absent radius syndrome (TAR). Marjorie and Joe's parents do not have any of these symptoms.

**SOURCE:** OMIM 274000

**WORKSHEET:**

1. Draw a pedigree for this family.




2. Does the fact that different individuals have different symptoms indicate that this condition is pleiotropic, or a phenocopy?



3. The most likely mode of inheritance is _____

because _____ .


4. If Pete marries a woman who is homozygous for the wild type allele of the gene that causes this condition, the probability that a child of theirs would be a carrier is _____.

23

# ANHIDROTIC ECTODERMAL DYSPLASIA                CHAPTER 6

In 1875, Charles Darwin described a family with anhidrotic ectodermal dysplasia:

"Ten men, in the course of 4 generations, were furnished, in both jaws taken together, with only 4 small and weak incisor teeth and with 8 posterior molars. The men thus affected have very little hair on the body, and become bald early in life ... It is remarkable that no instance has occurred of a daughter being affected ... Although the daughters are never affected, they transmit the tendency to their sons; and no case has occurred of a son transmitting it to his sons."

In additional families with the disorder, some females were very mildly affected, with small or malformed teeth and abnormal sweat glands. The lack of hair and sweat glands was patchy, with only some areas of skin affected. A test to detect female carriers reveals patchy distribution of sweat glands on the back.

**SOURCES:** OMIM 305100, 224900

**WORKSHEET:**

1. The most likely mode of inheritance for this condition is _____ _____.

2. Is the phenotype sex limited? Why or why not?


3. Gene expression in females is "patchy" because of _____.


4. The probability that a son of a woman who has patchy symptoms inherits anhidrotic ectodermal dysplasia is _____.


5. For a woman to have a case as severe as an affected male, possible phenotypes of her parents would be _____

_____.

# BLUE DIAPER SYNDROME — CHAPTER 6

Gloria adores her nephews Will and Charlie, who have an unusual inherited condition. Because of abnormal transport of the amino acid tryptophan across the small intestinal lining, bacteria act on urine precursors to produce a compound that turns indigo blue upon contact with the air. Will's wet diapers turn blue! By the time the couple had Charlie, they were accustomed to the blue urine.

The family doctor assured Gloria and her nephews' parents, Gloria's sister Edith and her husband Archie, that other relatives needn't worry about recurrence. She explained that it is autosomal recessive and affected Will and Charlie because Edith and Archie are carriers. The condition hadn't appeared in previous generations, and likely wouldn't again, unless Edith and Archie had more children, each of whom faced a 25 percent chance of inheriting the trait. So Gloria and her husband Michael are quite surprised when their fraternal twins Marshall and Joey each produced blue diapers!

**SOURCE:** OMIM 211000

**WORKSHEET:**

1. The error that the physician made was that _____

_____.

2. Draw a pedigree for this extended family, indicating their unusual disorder.

3. Gloria and Edith wonder if their grandsons will inherit blue diaper syndrome. Will they? Why or why not?

4. The blue diaper phenotype is likely to be darker after the baby eats food containing a lot of _____.

# CHRONIC GRANULOMATOUS DISEASE                                    CHAPTER 6

A 67-year-old man is hospitalized with fever, chills, headache, and lack of appetite due to infection by the bacterium *Pseudomonas cepacia*. This rare infection also killed the man's 5-year-old grandson four years earlier. The grandson had chronic granulomatous disease, an X-linked recessive condition in which certain white blood cells can engulf bacteria, but cannot then destroy them as white blood cells normally would. The grandfather apparently had the condition too, although he had been healthy until the present infection. Most people with this disorder have frequent bacterial infections.

Healthy white blood cells called phagocytes kill engulfed bacteria by producing a chemical called superoxide, which is formed when oxygen picks up electrons. The electrons are passed by a complex of four proteins, each of which is encoded by a different gene. Chronic granulomatous disease results from an abnormality in one of these proteins whose gene, gp91-phox (phox stands for phagocyte oxidase) resides on the X chromosome.

When the family learns that the same illness that killed the grandson is causing the grandfather's infection, other relatives are tested, specifically the females, to see if they are carriers. This is easy to do – a blue dye, tetrazolium, changes color if superoxide is present. The grandfather's daughter and her two daughters (by her second husband) have some cells that produce superoxide and some that do not.

The pedigree for this family is:

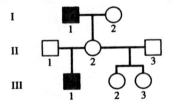

SOURCE: OMIM 306400

WORKSHEET:

1. The grandfather's daughter (individual II2) is pregnant. She is very worried that a son would be ill, as her first son was, but her present husband (individual II3) assures her that is not possible because he is the father now. Is the husband correct?

2. Was the tetrazolium test for carrier status necessary for the grandfather's daughter? Cite a reason for your answer.

3. White blood cells are obtained from the fetus, and they lack superoxide activity. Further genetic information that might be useful in evaluating the fetus' future is _____.

4. How does the environment influence the expression of the gp91-phox gene?

5. How would the pedigree differ if the disease is inherited as an autosomal recessive trait?

# CONGENITAL MUSCULAR DYSTROPHIES     CHAPTER 6

Muscular dystrophies cause progressive muscle weakness. The most common types of muscular dystrophy (Duchenne and Becker) are X-linked and affect the muscle protein dystrophin. Although accounting for only a small amount of the proteins in a muscle cell, dystrophin is very important, because it binds to the inner face of the cell membrane and provides strength that enables the cell to withstand the force of contraction.

Dystrophin must bind to several dystrophin-associated proteins in order to function. Several autosomal recessive types of muscular dystrophy affect different proteins, and are called congenital muscular dystrophies. Some affect collagen, the connective tissue protein, and some affect integrins, which are proteins involved in cellular adhesion. Forty percent of congenital muscular dystrophies affect laminin alpha 2, a protein that binds a protein called alpha dystroglycan, which in turn is a dystrophin-associated protein.

**SOURCES:** OMIM 156225

Martin Brockington et al. December 2001. Mutations in the fukutin-related protein gene (FKRP) cause a form of congenital muscular dystrophy with secondary laminin alpha 2 deficiency and abnormal glycosylation of alpha-dystroglycan. *The American Journal of Human Genetics* 69:1198-1209.

**WORKSHEET:**

1. What is the physical basis of the similarity of symptoms of laminin alpha 2 muscular dystrophy and Duchenne muscular dystrophy?




2. Why is the risk of a person who is a carrier for congenital muscular dystrophy due to a mutation in an integrin gene having an affected child with a carrier of congenital muscular dystrophy due to a mutation in a collagen gene not 25 percent?

3. The mothers of the affected children in the pedigrees below each have a newly married sister. Russell Rivington has Duchenne muscular dystrophy. Martha O'Toole has a type of congenital muscular dystrophy. What is the risk that Cheryl and Rob have a child (boy or girl) with muscular dystrophy? What is the risk that Megan and Tom have an affected child? Why do these risks differ?

The Rivingtons

The O'Tooles

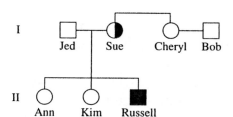

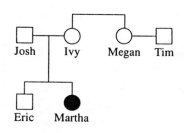

Duchenne muscular dystrophy

Congenital muscular dystrophy

4. The risk that Kim has a son with DMD is _____.

# INTERSEX  CHAPTER 6

An individual who has ambiguous genitalia, or genital structures characteristic of both sexes, is medically considered to be intersex. Athletes who are intersexes sometimes have difficulty in determining whether they should compete as male or female. Below are some real cases.

**SOURCE:** http://www.geocities.com/Colosseum/Pressbox/6031/hr.html

**WORKSHEET:**

1. Zdenka Koubkowa was born in Czechoslovakia in 1913, and competed in the women's 800-meter race in 1934 in London, breaking a world record. She was born with ambiguous genitalia, and was raised as a girl. However, she always felt that she was really a man, and later in life, cut her hair and began dressing and living like a man. Which medical condition described in Chapter 6 was a likely diagnosis for Koubkowa?

2. Maria Jose Martinez Patino was a Spanish hurdler. Although she had female genitalia, in 1986 chromosome tests revealed that she was XY, and she was banned from further athletic competition as a female. Her lack of menstruation went unnoticed among elite athletes, who often cease menstruating if body fat is too low. Which medical condition described in Chapter 6 was a likely diagnosis for Patino?

3. Debbie Meyer swam in the 200 meter freestyle event in the 1968 Olympics in Mexico. She paid for a chromosome test before competing, however, because she feared that her intense exercise had made her become an XY. Is this possible?

4. Describe a chromosomal aberration that would produce an individual with a female phenotype with an XY chromosome constitution.

# CLEFT LIP WITH OR WITHOUT CLEFT PALATE  CHAPTER 7

Cleft lip (failure of the upper lip to fuse) and cleft palate (failure of the soft palate to fuse) are multifactorial conditions that occur during prenatal development. More than 200 syndromes include cleft lip with or without cleft palate.

When clefts occur alone, heritability ranges from 77% to 97%. Individuals with an affected relative have an increased risk of being born with a cleft. The risk of recurrence depends upon the degree of relationship to an affected individual as follows:

- if one parent has a cleft, risk to a child is 4%
- if one sibling has a cleft, risk to other sibling is 4%
- if two siblings have a cleft, risk to third sibling is 9%
- if one parent and one sibling have a cleft, risk to another sibling is 17%

**SOURCE:** OMIM 119530

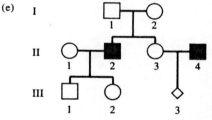

**WORKSHEET:**

1. In each family depicted in (a) through (e), a couple expecting a child is concerned that she or he will have a cleft lip with or without cleft palate. For each couple, cite the risk that the fetus (designated as a diamond) faces:

   a.

   b.

   c.

   d.

   e.

2. An estimate of the likelihood of recurrence based on population statistics for similar individuals is called the _____ risk.

31

3. Heritability is an estimate of _____.

4. Would you expect both members of MZ twin pairs to be more likely to each have cleft lip and cleft palate than both members of DZ twin pairs, considering twins in which at least one individual is affected? Cite a reason for your answer.

5. Cleft lip and cleft palate can also be caused by exposure to certain drugs during embryonic development. This is an example of

    a. pleiotropy
    b. penetrance
    c. a phenocopy
    d. a phenotype
    e. phenylketonuria

# COMPLEX TRAITS AMONG THE HUTTERITES — CHAPTER 7

The Hutterites today number more than 35,000 and live on 350 communal farms, called colonies, in the northern U.S. and western Canada. They descend from a religious sect that lived in the Tyrolean Alps in the 1500s. In the 1870s, 900 of them moved to what is now South Dakota, settling on three large farms. They had many children, did not use birth control, and prepared and ate their meals in communal dining halls, all living essentially the same life. Genealogists have traced today's Hutterites to 90 ancestors from the 1700s.

Carole Ober and her co-workers at the University of Chicago study a group of Hutterites in South Dakota who descend from 64 ancestors. They number 722 individuals who live in 9 colonies, sharing diet and a healthy lifestyle that includes exercise and forbids smoking.

The researchers determined broad and narrow heritabilities for several multifactorial traits in this population. Heritability is the proportion of variance in a trait that can be attributed to genes. Narrow heritability (h) includes only the input of recessive alleles, but broad heritability (H) also takes into account rare dominant alleles that contribute to the trait, as well as epistasis (interactions between genes, as opposed to additive effects). If broad and narrow heritabilities are equal, then there are no dominant alleles or epistasis contributing to the trait. The data are as follows:

| Trait | Broad (H) | Narrow (h) | H in other populations |
|---|---|---|---|
| LDL | .96 | .36 | .34 - .50 |
| HDL | .63 | .63 | .42 - .83 |
| BMI | .54 | .54 | .32 - .59 |
| Serotonin levels | .99 | .52 | no data |
| Lung function | .40 | .40 | .06 - .45 |
| IgE | .63 | .63 | .47 - .79 |
| Height | .83 | .83 | .66 - .78 |
| Diastolic blood pressure | .21 | .21 | .28 - .44 |
| Systolic blood pressure | .45 | 0 | .18 - .54 |
| Triglycerides | .37 | .37 | .19 - .55 |
| Eosinophilia | .32 | .32 | .30 |

Explanation of traits:
HDL and LDL are lipoproteins that carry cholesterol. Low LDL and high HDL are associated with a healthy cardiovascular system.

Diastolic blood pressure is the blood pressure when the heart is relaxed, and systolic blood pressure occurs when it is contracting.

Eosinophilia is a measure of eosinophils, a type of white blood cell.

IgE is a type of antibody associated with allergies such as asthma.

Serotonin is a neurotransmitter associated with mood, level of arousal, and other characteristics.

BMI is body mass index, a measure of weight that considers height.

Triglycerides are a type of lipid (fat).

**SOURCES:** Complex Trait Mapping in the Hutterites, Carole Ober
http://www.genes.uchicago/edu/fri/oberrsh.html

Carole Ober et al. November 2001. The genetic dissection of complex traits in a Founder population. *The American Journal of Human Genetics 69:1068-1079.*

**WORKSHEET:**

1. Two reasons that the Hutterites are particularly useful to study multifactorial traits are_____

_____

and _____.


2. The traits for which dominant alleles contribute are _____

_____ .


3. What information can be gained by comparing the broad heritability among the Hutterities for a particular trait to the value seen in other populations?

# ALCOHOLISM          CHAPTER 8

Liver cells metabolize ethanol in two enzyme-catalyzed steps:

$$\text{ethanol} \xrightarrow{\text{alcohol dehydrogenase (ADH)}} \text{acetaldehyde} \xrightarrow{\text{aldehyde dehydrogenase (ALDH)}} \text{acetic acid}$$

Buildup of acetaldehyde causes the unpleasant symptoms associated with a hangover – flushing, palpitations, headache, nausea and vomiting. Most people produce enough aldehyde dehydrogenase (ALDH) to rapidly convert acetaldehyde into acetic acid, so that it takes many drinks to bring on symptoms. However, in Japan and China, about half of the population has at least one mutation in the ALDH gene. A study of 643 Japanese men found that those homozygous for the ALDH mutation are the least likely to drink alcohol because doing so makes them feel very ill very fast. The heaviest drinkers were homozygous for the wild-type form of the ALDH gene. Heterozygotes had varied responses.

Another study showed that women with alcoholism who have a variant of ADH that speeds conversion of ethanol to acetaldehyde are less likely to have children who suffer from fetal alcohol syndrome (FAS) than women with alcoholism who do not have this enzyme variant. Another variant of the ADH gene is more common among African-Americans and some native American groups than others.

Other studies seek polygenic explanations for differences in intoxication rate. Researchers scanned the genomes of 1,004 individuals for 1,494 single nucleotide polymorphisms (SNPs), and identified 8 areas of the genome that are associated with alcohol or nicotine abuse. Finally, at the population level, incidence of FAS varies greatly, which may reflect genetic differences in alcohol metabolism:

| Population | Incidence of FAS (per 1000 births) |
| --- | --- |
| General U.S. | 0.33 – 2.2 |
| Native Americans | 8 |
| Inner city African Americans | 2.29 |
| Western Cape population of South Africa | 40 – 70 |

**SOURCES:** OMIM 103720

George R. Uhl et al. December 2001. Polysubstance abuse vulnerability genes: genome scans for association, using 1,004 subjects and 1,494 single nucleotide polymorphisms. *The American Journal of Human Genetics* 69:1290-1300.

**WORKSHEET:**

1. How might the fast-acting variant of ADH protect against FAS?




2. Why do the Japanese and Chinese men who are homozygous for the ALDH mutation become ill faster than other men?




3. How can the tendency to develop alcoholism be Mendelian, polygenic, and multifactorial?




4. What would be the use, if any, of using microarrays of SNPs to determine how likely people are to develop alcoholism?




5. Do you think that population-based screening for genes that can affect alcoholism should be restricted to groups of people who are more likely to have predisposing genotypes – or would such a practice be discriminatory? If it were instituted, which groups might obtain useful information from undergoing which tests?

# PART THREE

# DNA AND CHROMOSOMES

**CHAPTER 9**        **DNA STRUCTURE AND REPLICATION**
                             DNA replication

**CHAPTER 10**      **GENE ACTION AND EXPRESSION**
                             Alpha$_1$-antitrypsin deficiency
                             Hypoxia-inducible factor 1

**CHAPTER 11**      **GENE MUTATION**
                             Bloom syndrome
                             DNA repair
                             Gyrate atrophy
                             Open-angle glaucoma
                             Otospondylomegaepiphyseal dysplasia
                             Tay-Sachs disease
                             Von Willebrand disease

**CHAPTER 12**      **CHROMOSOMES**
                             DiGeorge syndrome
                             Down syndrome
                             Tetrasomy 12p
                             Turner syndrome
                             Williams syndrome

# DNA REPLICATION — CHAPTER 9

Watson and Crick's idea that the double helix separates and new bases come in to create two double helices, each with one old and one new strand, was envisioned at the DNA level. At the same time that Meselson and Stahl were demonstrating the semi-conservative nature of DNA replication and disproving the conservative and dispersive hypotheses, other researchers were exploring DNA replication at the level of the chromosome. They used the fact that chromosomes absorb dyes – hence the name chromosome, which means "colored body."

Taylor Herbert, Philip Woods, and Walter Hughes, at Brookhaven National Laboratory, explored DNA replication using the Easter lily in the 1950s. They labeled newly formed DNA with radioactive thymidine, a precursor of thymine. When they overlaid the treated DNA with photographic film, the radioactivity exposed the film, yielding dots that indicated the sites of replicating DNA. If a cell was allowed to divide and then the DNA was labeled, no dots formed, because the label was not incorporated – it was too late. If the cell divided once in the presence of the radioactive label and was then examined after the DNA had replicated again, both sister chromatids of each chromosome were labeled. If a cell divided once in the presence of the label, then was removed and allowed to divide again without the label, then one sister (half) of each replicated chromosome was labeled.

**SOURCE:** Taylor, J. Herbert. 1958. The duplication of chromosomes. *Scientific American*.

**WORKSHEET:**

1. How would the results of this experiment have differed if DNA replication were conservative? Dispersive?

2. The researchers interpreted their results to mean that chromosomes are composed of DNA. But they were puzzled. How could a molecule that is more than a yard long, with some 300 million twists, fold into a short, thick chromatid? Explain how this is indeed possible.

3. The researchers also knew that chromosomes were composed of DNA as well as protein. They suggested two possible structures for chromosomes. In the ribbon model, DNA emanates in short branches from a protein column, the branches waving freely and not connected to each other. In the two-column model, DNA chains zigzag between two protein columns, which can contort as the genetic material replicates. The zigzags are continuous. Which model is more consistent with DNA's true structure, and why?

# ALPHA$_1$-ANTITRYPSIN DEFICIENCY — CHAPTER 10

Jamal developed difficulty breathing shortly after birth, and had an unusually severe case of newborn jaundice, indicating a liver problem.

Blood tests revealed that he had the autosomal recessive condition alpha$_1$-antitrypsin (AAT) deficiency. AAT is an enzyme that normally dismantles another enzyme, elastase. White blood cells in the lungs produce elastase to destroy infectious bacteria. Without enough AAT, elastase builds up and destroys more than bacteria – it eats away at lung tissue, too, causing emphysema. Several dozen alleles of the AAT gene are known, but a particularly severe one is called "Z". A Z allele has mutations at two sites in the gene. Jamal is a "ZZ", a homozygote – and as a result has the most serious form of AAT deficiency.

Jamal's liver cells could manufacture AAT, but the amino acid chains could not fold properly. Instead, they stuck to each other, forming structures called inclusions that built up in the endoplasmic reticulum, rather than being secreted. Over time, the build up hardened his liver – this is cirrhosis, the same problem that can result from drinking too much alcohol. He will likely die by middle age.

**SOURCE:** OMIM 107400

Carrell, Robin W. and David A. Lomas. January 3, 2002. Alpha$_1$-antitrypsin deficiency – a model for conformational diseases. *The New England Journal of Medicine*, vol. 346(1):45-53.

**WORKSHEET:**

1. Explain how AAT deficiency is both an inborn error of metabolism and a conformational disease.

2. Two ways that Jamal can alter his lifestyle once he gets older to possibly extend his life are _____ and _____ .

3. The defect in Jamal's case of AAT is at the level of

   a. DNA replication
   b. Transcription
   c. Translation
   d. None of the above


4. In what way is this form of AAT deficiency similar to Alzheimer disease, Huntington disease, and the transmissible spongiform encephalopathies (see chapter 2)?

# HYPOXIA-INDUCIBLE FACTOR 1        CHAPTER 10

If a blood clot lodged in a coronary artery cuts off the oxygen supply to the heart muscle for more than a few seconds, the lack of oxygen activates transcription of the hypoxia-inducible factor 1 (HIF-1) gene in cardiac muscle cells. The gene's product, a transcription factor, activates other genes that:

- induce angiogenesis (formation of new blood vessels)
- promote glycolysis (provides energy in the absence of oxygen)
- stimulate production of the hormone erythropoietin (EPO), which prompts the bone marrow to release red blood cell precursors

HIF-1 has two parts. One part, HIF-1α, is an 826 amino acid long peptide that activates genes that promote the effects listed above. The second part of the molecule, HIF-1B, is a nuclear translocation protein that transports HIF-1α out of the nucleus. HIF-1B also associates with various other peptides to form other types of two-part proteins, or heterodimers. Based on tests that demonstrated production of HIF-1α potein in various human cell types grown in culture and deprived of oxygen, researchers are developing detection of HIF-1α as a test to confirm early damage from a heart attack.

**SOURCES:** OMIM 193300

Lee, Sang H. et al, March 2, 2000. Early expression of angiogenesis factors in acute myocardial ischemia and infarction. *The New England Journal of Medicine* 342:626.

**WORKSHEET:**

1. The mRNA that encodes HIF-1α is _____ bases long.


2. In vitro tests measure the amount of mRNA and not tRNA or rRNA because




3. How can activation of a transcription factor exert several effects?

4. Why could a mutation in either of two different genes affect production of HIF-1 protein?




5. What would be the phenotype associated with a mutation in the HIF-1 gene?

# BLOOM SYNDROME   CHAPTER 11

Bloom syndrome causes increased risk of cancer, particularly leukemia; facial rash in response to sun exposure; and severely impaired immunity. The oldest known person with the syndrome lived until middle age, but most affected individuals die as children or adolescents. Of the 100 people in the world who have the autosomal recessive Bloom syndrome, 24 have parents who are cousins. Many of the affected families come from isolated communities. Bloom syndrome is caused by a mutation in the gene that encodes ligase 1, which is an enzyme that connects the sugar-phosphate backbone in newly replicated DNA.

At the cellular level, signs of Bloom syndrome include:

- DNA replication forks that progress very slowly
- A delay in joining of Okazaki fragments into larger DNA molecules
- Slow growth of cells in culture (slowed cell cycle)
- Increased frequency of chromosome breakage and rearrangement
- Increased crossing over between chromatids of replicated chromosomes
- Lack of p53 protein accumulation after exposure to ultraviolet (UV) radiation.
- Increased sensitivity to DNA damaging agents

*E. coli* and yeast cells with mutant ligases have similar characteristics to cultured cells from people who have Bloom syndrome.

**SOURCE:** OMIM 210900

**WORKSHEET:**

1. How does the cellular phenotype explain the whole body phenotype?




2. The symptom that indicates a defect in DNA repair is

_____ .

3. Is the defect more likely in nucleotide excision repair or base excision repair? Cite a reason for your answer.

4. How could you demonstrate whether the chromosome abnormalities that are associated with Bloom syndrome are a cause or effect of the illness?

5. How does the environment affect expression of Bloom syndrome?

# DNA REPAIR  CHAPTER 11

When skin is damaged by ultraviolet (UV) radiation, pyrimidine dimers form, and they are either repaired, or the cell dies by apoptosis and is shed, in the form of skin peeling. Researchers are examining various biochemicals that might be added to sunblock products to lessen the harm from sunburn, without abolishing its protective function. In one experiment, researchers exposed mice to UV radiation with or without a cream containing interleukin-12, a type of immune system protein called a cytokine. The treated mice had fewer thymine dimers and fewer peeling skin cells than did the mice that did not receive interleukin-12.

**SOURCE**: Agatha Schwartz et al. January 2002. Interleukin-12 suppresses ultraviolet radiation-induced apoptosis by inducing DNA repair. *Nature Cell Biology* 2:1.

**WORKSHEET:**

1. Two conclusions that would explain these observations are:

2. Before interleukin-12 is added to sunscreens products for human use, what further experiments should be done?

3. If DNA repair occurs, would it more likely be nucleotide excision repair or base excision repair?

# GYRATE ATROPHY
CHAPTER 11

Gyrate atrophy is a degeneration of the retina that begins in late adolescence as night blindness and progresses to blindness. The cause is mutation in the gene that encodes a mitochondrial enzyme, ornithine aminotransferase (OAT), on chromosome 10. This autosomal recessive inborn error of metabolism causes build up in body fluids of ornithine, which derives from the amino acid arginine, found in dietary protein. Researchers sequenced the OAT gene for five patients with the following results:

- Patient A: A change in codon 209 of UAU to UAA
- Patient B: A change in codon 299 of UAC to UAG
- Patient C: A change in codon 426 of CGA to UGA
- Patient D: A 2 base deletion at codons 64 and 65 that results in a UGA codon at position 79
- Patient E: Exon 6, including 1,071 bases, is entirely deleted.

**SOURCE:** OMIM 258870

**WORKSHEET:**

1. Patient _____ has both a frameshift and a nonsense mutation.

2. Patients A, B and C have in common _____.

3. Another patient, F, has the mutations seen in patients A and B. How is this possible?


4. Gyrate atrophy does not exhibit maternal (mitochondrial) inheritance because

_____ .

5. Patient E is missing _____ amino acids.

6. Suggest a way to relieve or slow the symptoms of gyrate atrophy.

# OPEN-ANGLE GLAUCOMA    CHAPTER 11

In open-angle glaucoma, cell death in the optic nerve prevents fluid from draining out of the eye, which causes pressure to build up in the eyeball. This interferes with peripheral vision, and can be relieved with surgery, laser treatment, or drugs. Vision is usually not noticeably affected until after age 40. A gene that causes this condition is on chromosome 1, and is called GLC1A. Many mutations are known to cause this condition, but the gene also has polymorphisms that do not affect the phenotype.

**SOURCES:** OMIM 137750, 602432

Alward, Wallace L.M. et al. April 9, 1998. Clinical features associated with mutations in the chromosome 1 open-angle glaucoma gene (GLC1A). *The New England Journal of Medicine* 338:1022.

**WORKSHEET:**

1. Mutations below in the GLC1A gene that cause open-angle glaucoma. Write codon changes that could account for these changes in amino acid sequence:

    a. tryptophan to arginine

    b. glycine to valine

    c. tyrosine to histidine

2. The following polymorphisms in the GLC1A gene do not alter the amino acid sequences. Write in possible codon changes:

    a. proline to proline

    b. threonine to threonine

    c. glutamate to glutamate

3. How can a change in a gene's sequence not alter the protein product?

4. A gene that when mutant can cause open-angle glaucoma might encode a protein that participates in

    a. chromosome segregation
    b. apoptosis
    c. meiosis
    d. the polymerase chain reaction

# OTOSPONDYLOMEGAEPIPHYSEAL DYSPLASIA        CHAPTER 11

"OSMED" is an autosomal recessive disorder whose symptoms arise from the absence of the alpha chains of collagen. Symptoms include short stature, severe hearing loss, underdeveloped facial features and cleft palate, enlarged joints, a bent spine, and nearsightedness. Of 10 families examined, four were consanguineous, and mutations found to be either deletions or nonsense. In the consanguineous families, the affected individuals were homozygotes for the mutant allele. In the other families, however, the affected individuals had two different mutant alleles.

**SOURCES:** OMIM 214150

Melkoniemi, Mia et al. February 2000. An autosomal recessive disorder otospondylomegaepiphyseal dysplasia is associated with loss-of-function mutations in the COL11A2 gene. *The American Journal of Human Genetics* 66:368.

**WORKSHEET:**

1. Why is OSMED an autosomal recessive disorder, but Ehlers-Danlos syndrome type IV (see the case for this disorder), which affects the same protein, is an autosomal dominant condition?

2. The basis for the pleiotropy associated with OSMED is that ___

_____.

3. Why are the affected people in the consanguineous families homozygotes, whereas those in the other families are heterozygotes?

4. A nonsense mutation and a deletion mutation can each cause absence of the gene product because _____

_____.

# TAY-SACHS DISEASE  CHAPTER 11

Tay-Sachs disease is an autosomal recessive condition that causes progressive nervous system degeneration. A child is deaf and blind by one or two years and usually dies by age three. The disorder is very rare in the U.S. because of screening programs to identify carriers. Since the 1970s, most couples who are both identified as carriers have chosen to avoid the birth of affected children because there is no treatment. Worldwide, Tay-Sachs disease has been most prevalent among Ashkenazi Jews, French Canadians, the Pennsylvania Dutch, Cajuns, and Moroccan Jews. Today, the few cases reported each year generally appear in other population groups that aren't screened.

More than 30 mutations are known in the HEXA gene on chromosome 15 that cause Tay-Sachs disease. The wild type allele encodes the alpha subunit of hexosaminidase, a dimeric (two-part) enzyme that also has a beta subunit, encoded by a different gene.

The five people below have different HEXA mutations:

- Patient A has a 4 base pair insertion in an exon.

- Patient B has a G to C mutation in the splice site of intron 12, which creates a recognition site for the restriction enzyme Dde1.

- Patient C has a deletion that eliminates phenylalanine (*phe*) from the protein at amino acid position 304. Position 305 is also a phenylalanine.

- Patient D has a C to G mutation at amino acid 180, which alters a UAC to a UAG in the mRNA.

- Patient E has a G to A mutation at amino acid position 170, which alters an arginine (*arg*) to a glutamine (*gln*).

**SOURCE:** OMIM 205400

**WORKSHEET:**

1. Patient D's hexosaminidase A is abnormal in that it _____

_____.

2. Two codon changes that could account for patient E's substitution of a *gln* for an *arg*, involving a G to A transition mutation, are ____ to _____ and ____ to _____.

3. The number of DNA bases deleted in patient C is _____.

4. How might patient B's mutation alter the gene product?

5. How might another genetic condition cause symptoms similar to those of Tay-Sachs disease?

6. Patient _____ has an altered reading frame for the HEXA gene.

# VON WILLEBRAND DISEASE     CHAPTER 11

Von Willebrand disease is the most common inherited bleeding disorder, and it is transmitted as an autosomal recessive trait. Eighty percent of sufferers have a mild form of the illness, with prolonged bleeding from a wound, easy bruising, and very heavy menstrual periods. About 1% of sufferers have a very severe form. Clotting factors are given to control the symptoms.

A mutation in the gene that encodes von Willebrand factor (vWF) causes the disease. The encoded protein is a large, multi-subunit plasma glycoprotein that interacts with platelets and clotting factor VIII (the protein absent in hemophilia A) to control blood clotting. People with severe von Willebrand disease lack vWF. The gene maps to chromosome 12p. It spans 178 kilobases, includes 52 exons, and is transcribed into an mRNA of 8.8 kilobases. The gene product is translated into a precursor form with only the mature protein forming vWF as follows:

| signal peptide | propeptide | mature protein |
|---|---|---|
| 22 amino acids | 741 amino acids | 2050 amino acids |

Swedish researchers sequenced the vWF gene from 25 patients with the severe form of the illness. All had a point mutation that alters a CGA mRNA codon to a UGA codon. One patient had this particular mutation in an intron, where it introduces a new recognition site for the restriction enzyme Dde1.

**SOURCE:** OMIM 193400

**WORKSHEET:**

1. The number of DNA bases that are not represented in the mature protein product of the vWF gene is _____.

2. The number of DNA bases that encode the form of vWF found in the bloodstream is _____.

3. The point mutation identified in the 25 von Willebrand disease patients is a _____ and a _____ mutation.

    a. transversion, nonsense
    b. transversion, missense
    c. transition, nonsense
    d. transition, missense

4. The dinucleotide CG is a mutational hot spot, meaning that it is implicated in mutations causing a variety of disorders. Why might this dinucleotide be particularly prone to mutations that drastically alter the phenotype?




5. A mutation adds a restriction site for Dde1. Does a restriction enzyme digest result in pieces of DNA that are larger or smaller than for the wild type allele? Explain your answer.




6. There is also an X-linked form of von Willebrand disease. How does the structure and function of vWF explain this genetic heterogeneity?

# DIGEORGE SYNDROME  CHAPTER 12

The first two generations of the Findley family Shown here are healthy, so they are surprised when Laura and her husband Aaron, and Dylan and his wife Iris, have children.

Karen, Leslie, and Lance have a rare condition called DiGeorge syndrome. The children have small facial features and low, rotated ears. More serious symptoms are defects in the blood vessels leading from the heart, an underdeveloped thymus gland (impairing immunity), and under-developed parathyroid glands (disrupting calcium metabolism.)

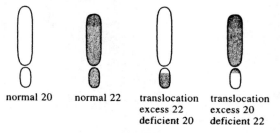

Dylan and Iris consult a genetic counselor, Ms. Reese, who is concerned that their niece Karen has the same syndrome as Leslie and Lance. She learns that Iris has had several spontaneous abortions, as has her sister-in-law Laura. Noting the combination of repeated birth defects and pregnancy losses, Ms. Reese suggests that Dylan and Laura, their siblings, children and parents be karyotyped. The results are, unfortunately, as the genetic counselor suspects.

Perry, Holly, Ariel, Sherri, and Shane have normal chromosomes. Laura, Dylan, Zach, and Kim have a reciprocal translocation between chromosomes 20 and 22. They have no symptoms but are translocation carriers with the following chromosomes:

normal 20    normal 22    translocation excess 22 deficient 20    translocation excess 20 deficient 22

The children with DiGeorge syndrome – Karen, Leslie, and Lance – have partial monosomy 22 and partial trisomy 20, a genetic imbalance responsible for their symptoms. The reverse situation – partial trisomy 22 and partial monosomy 20 – probably accounts for Laura's and Iris' spontaneous abortions because these imbalances halt development.

**SOURCE:** OMIM 188400

**WORKSHEET:**

1. Zach is an only child. His mother had three spontaneous abortions before giving birth to him, then she had a baby that died shortly after birth, which Zach barely remembers. Diagram the distribution of chromosomes 20 and 22 in meiosis and fertilization to depict how these pregnancy problems arose.

2. Why did Iris have reproductive difficulties if she is not a blood relative of Zach and has normal chromosomes?

3. Karen, Leslie, and Lance have

    a. two normal copies of chromosomes 20 and 22
    b. two normal chromosome 20s, one normal chromosome 22, and one chromosome 22 with some chromosome 20 material replacing some chromosome 22 material
    c. two normal chromosome 22s, one normal chromosome 20, and one chromosome 20 with some chromosome 22 material replacing some chromosome 20 material.
    d. One normal chromosome 20, one normal chromosome 22, and two translocated chromosomes

4. To be a translocation carrier like Zach, Laura, Dylan and Kim, one must inherit _____ of the translocated chromosomes.
    a. one
    b. both
    c. neither

# DOWN SYNDROME                                            CHAPTER 12

Kaneesha is 26 years old, and her husband Marshall is 28. When Kaneesha becomes pregnant, she asks her obstetrician about prenatal tests, because a cousin had a child with Down syndrome at age 27, and her two sisters have each had two miscarriages. She tells the doctor of this family history. This is Kaneesha's first pregnancy, although she and Marshall have been trying to become pregnant for several years.

The doctor advises Kaneesha to have a maternal serum marker (blood) test, called the "triple test," instead of amniocentesis, because she is under age 35. The triple test is so-called because it measures levels of three substances – alpha fetoprotein, human chorionic gonadotropin, and a form of estrogen. The doctor tells the couple that the test is only a screen, and can detect up to 60 % of cases of Down syndrome that are caused by an extra chromosome 21. When Kaneesha gives birth to Nathaniel, it is clear that he has Down syndrome, but his karyotype does not show a trisomy. Kaneesha and Marshall sue the obstetrician, on the grounds that they were advised to take the wrong prenatal test.

## WORKSHEET:

1. Why did the doctor advise Kaneesha not to have amniocentesis – and why should she have had the test?

2. How can Nathaniel have Down syndrome if his cells have 46 chromosomes?

3. What is the risk of Down syndrome in future children for this couple?

# TETRASOMY 12p                                    CHAPTER 12

People with tetrasomy 12p, also known as Pallister-Killian syndrome, have an unusual combination of symptoms:

- Alopecia (hair loss)
- Mental retardation
- Seizures
- Extra nipples
- Defects in the diaphragm muscle
- Slow movements

The face has very coarse features, such as a large mouth, wide set eyes, puffed cheeks and a small nose. Most individuals die as infants, but a few survive into adolescence. Karyotypes reveal that affected individuals have an isochromosome that consists of material from the short arm of chromosome 12, but only in some of their cells. The remainder of their cells have normal karyotypes. The condition is also called tetrasomy 12p.

**SOURCE:** OMIM 601803

**WORKSHEET:**

1. Why are all individuals with this syndrome mosaics for the isochromosome?

2. Explain or illustrate how an isochromosome forms.

3. All cases are sporadic – that is, none of the affected individuals has a relative who also has the condition. Why is this so?

# TURNER SYNDROME                                    CHAPTER 12

Most people who have Turner syndrome have one X chromosome and no Y chromosome in each somatic cell. About 15% of Turner patients have a different chromosomal anomaly, an isochromosome for the X, in which a single chromosome consists of two long arms but no short arm. A smaller percentage of patients have yet another aberration that occurs in only some of their cells: a small ring chromosome derived from X chromosome material, in addition to a normal X.

In a study at Henry Ford Hospital in Detroit, 190 Turner patients were evaluated cytogenetically. Five people were found to have an X-derived ring in some cells, but other cells were XO. All of these five patients were among the 6.3 % of Turner patients who are mentally retarded. The researchers hypothesized that the ring chromosome may cause mental retardation because it lacks genes that normally inactivate one X chromosome in each of a female's cells. They hypothesize that overexpression of some X linked genes causes the mental retardation.

## WORKSHEET:

1. The symptoms of Turner syndrome are:


2. What sex is a person who has Turner syndrome?

3. An isochromosome would be expected to cause symptoms for two reasons. They are:

   a.

   b.


4. One explanation for why no patients had the X-derived ring in every cell is that:



5. About 95% of Turner conceptions end as spontaneous abortions. Would you expect the frequency of Turner individuals with X-derived rings among spontaneous abortions to be higher or lower than that among Turner individuals who survive? Why or why not?

# WILLIAMS SYNDROME — CHAPTER 12

Williams syndrome is associated with a very unusual set of symptoms. Affected individuals have a characteristic face, including protruding ears, a small chin, and upturned nose. Their growth is usually delayed, and they often suffer from anxiety and heart problems. IQ is in the 50 to 60 range.

Most notable about some individuals with Williams syndrome is their uncanny musical ability. Many have absolute pitch – the rare ability to hear a note and be able to identify it. People who do not have Williams syndrome but who have inherited a perfect pitch gene display this skill, but only if they have musical training before age 6. Some people with Williams syndrome, however, can attain perfect pitch beyond this age. Musically gifted individuals with Williams syndrome also can retain incredible amounts of musical knowledge. One middle aged woman, for example, knows more than 2500 songs in 25 languages, and has sung classical music as well as recorded with the rock band Aerosmith. Her IQ is 55. A young man with Williams syndrome can tap out intricate rhythms, one with each hand, at the same time.

The Williams syndrome brain seems extra adept at written description, yet deficient in visualization. A person told to relate information about a hippo, for example, might draw a big blob, yet write an incredibly detailed and accurate description of the animal. At the same time, affected individuals cannot do the simplest calculations.

Most cases of Williams syndrome appear suddenly in a family. FISH analysis of many cases revealed that 95% of affected individuals have a *de novo* deletion of about 20 genes on chromosome 7, covering about 2 million DNA bases. The other 5% of affected individuals have an inversion of 1.5 million bases in the same region of chromosome 7 that is usually deleted. The DNA sequences that flank the ends of the inversion consist of many copies of short repeats.

Some people with only some of the signs and symptoms of Williams syndrome have microdeletions. FISH analysis indicates that a condition called supravalvular aortic stenosis, which causes blockages in arteries, is caused by deletion of a gene that encodes a form of the connective tissue protein elastin. This gene is in the same region that is deleted in Williams syndrome.


## SOURCES:

Brendan A. Maher. November 26, 2001. Music, the brain, and Williams syndrome. *The Scientist* 15(23):20-21.

Lucy R. Osborne et al. November 2001. A 1.5 million-base pair inversion polymorphism in families with Willaims-Beuren syndrome. *Nature Genetics* 29:321-325.

**WORKSHEET:**

1. Some health insurance agencies will only cover the costs of treating Williams syndrome if the patient has a deletion. Why might some families object to this limitation?

2. Suggest a mechanism related to the structure of chromosome 7 that can explain the sudden appearance of Williams syndrome in a family.

3. How can a deletion and an inversion cause the same syndrome?

4. Suggest a way that researchers can try to assign specific signs or symptoms to specific parts of chromosome 7.

5. How might studying Williams syndrome reveal information about normal learning?

# PART FOUR

# POPULATION GENETICS

**CHAPTER 13**  **WHEN ALLELE FREQUENCIES STAY CONSTANT**
　　　　　　　　The Ice Maiden

**CHAPTER 14**  **CHANGING ALLELE FREQUENCIES**
　　　　　　　　Type III 3-methylglutaconic aciduria
　　　　　　　　Ulnar-mammary syndrome

**CHAPTER 15**  **HUMAN ORIGINS AND EVOLUTION**
　　　　　　　　Novelty seeking and ADHD

# THE ICE MAIDEN                                              CHAPTER 13

The Ice Maiden was only 12 to 14 years old when Inca priests sacrificed her 500 years ago to their mountain gods. Anthropologists found her frozen remains in 1995, on Mount Ampato in the Peruvian Andes. Researchers extracted a tissue sample from her heart, and used PCR primers to amplify mitochondrial DNA sequences, then used polyacrilamide gel electrophoresis (PAGE) to visualize the amplified PCR products. The researchers recovered several pieces of DNA, in the 450-base range, and compared them to DNA sequences from several modern population groups. For one gene called HV1, databases located an exact match to one of the four founders of native Americans. A second gene, HV2, did not match native American founder populations, but was quite similar to a gene in the Ngobe people of Panama.

**SOURCE:** www.tigr.org/cet/edu/ice/

**WORKSHEET:**

1. Two advantages of tracing origins using mitochondrial DNA compared to using nuclear DNA are:




2. Was the population of which the Ice Maiden a member descended from native Americans, the Ngobe of Panama, or both? How do you know this?




3. Would the amplified DNA provide information on one or both of the Ice Maiden's parents?

# TYPE III 3-METHYLGLUTACONIC ACIDURIA              CHAPTER 14

The four types of 3-methylglutaconic aciduria (MGA) have "confused students for decades," wrote one researcher, because each variant has distinct symptoms, and each variant is caused by malfunction of a different protein. What the four types of the disorder share, however, is excretion of a substance called methylglutaconic acid in the urine. The variants of the disorder are:

| Type | Symptoms | Mode of inheritance |
|------|----------|---------------------|
| I    | Mild neurological impairment | Autosomal recessive |
| II   | Short stature<br>White blood cells in urine<br>Enlarged heart | X-linked recessive |
| III  | Early blindness<br>Uncontrollable movement<br>Spasticity<br>Poor balance<br>Impaired cognition | Autosomal recessive |
| IV   | Defects in heart, eyes, liver<br>nervous system, kidneys, | Autosomal recessive |

Type III MGA is of particular interest to geneticists because only 40 cases are known, and they are Jewish individuals of Iraqi descent, each with exactly the same symptoms. Each also has the same mutation in a gene called OPA3, which is on chromosome 19 and consists of two exons. The mutation changes a G to a C at an intron/exon splice site. As a result of this mutation, there is no mRNA for the gene product, which somehow results in buildup of MGA and spillover into the urine.

     Researchers detected the mutation in:

          10 affected individuals
          11 of their relatives who are carriers
          8 of 85 anonymous Israelis of Iraqi heritage

But they found no mutations among 55 North Americans.

     The researchers attribute the high prevalence of this mutation among Iraqi Jews to a founder effect. They trace the mutation to the original Middle Eastern gene pool of 12,000 Jews sent from Babylon in 586 B.C. after the First Temple was destroyed in Jerusalem. The population remained relatively isolated for 2,500 years, and there was some consanguinity. According to the most recent census (1998), 253,200 Iraqi Jews currently live in Israel. This group has high prevalence of other disorders that reflects a powerful founder effect, including a whopping 39% who have familial Mediterranean fever.

**SOURCES:** OMIM 258501

National Laboratory for the Genetics of the Israeli Population
http://www.tau.ac.il/medicine/NLGIP/nlgip.htm

Yair Anikster et al. December 2001. Type III 3-methylglutaconic aciduria (optic atrophy plus syndrome, or Costeff optic atrophy syndrome): identification of the OPA3 gene and its founder mutation in Iraqi Jews. *The American Journal of Human Genetics* 69:1218-1224.

**WORKSHEET:**

1. If 40 cases represent the initial 12,000 founders, the number of carriers of type III MGA in the current population is _____.

2. Cite two types of evidence that point to a founder effect accounting for the high prevalence of type III MGA among Iraqi Jews.

3. Two factors that could have sequestered this mutation in this population are:

4. Why were the four types of MGA considered a single disorder?

5. How would you design a screening test to detect milder cases of type III MGA?

# ULNAR-MAMMARY SYNDROME — CHAPTER 14

The symptoms of ulnar-mammary syndrome include underdeveloped nipples, short fingers and hands, short fourth toes, and lack of hair in the armpits. Affected males have very small penises and testes, a very low sperm count, and lack of libido (interest in sex). It is inherited as an autosomal dominant condition.

**SOURCE:** OMIM 181450

**WORKSHEET:**

1. How would natural selection have affected both sexes with this syndrome in the distant past?

2. In the distant past, why would most cases in females be due to new mutations, rather than inheriting the condition from an affected parent?

3. Although the symptoms of ulnar-mammary syndrome do not affect health, in terms of evolution this condition might be consider lethal. Why?

# NOVELTY SEEKING AND ADHD CHAPTER 15

Three percent of school-aged children in the U.S. today are diagnosed with attention deficit hyperactivity disorder (ADHD). They are hyperactive, impulsive, and have a short attention span. Many take medication to enable them to sit through a school day more comfortably. A generation ago, such children were often simply regarded as "behavior problems" and tolerated, or punished. Recent evidence, however, suggests that the ADHD phenotype is indeed a distinct characteristic, and that it may be prevalent today because in the past, it gave people a survival advantage.

Association and other studies have linked ADHD to variants of the DRD4 dopamine receptor gene, of which 56 alleles are known. In a recent screen for new variants among 600 people from all over the world, one new allele, called 7R, was found much more likely than can be attributed to chance among individuals who have signs of both ADHD and a behavior pattern called novelty seeking. Novelty seeking causes people to seek excitement and new experiences, but it is also correlated with a tendency to substance abuse. A look at children diagnosed with ADHD revealed that about half of them have this allele. The DNA sequence of the 7R allele indicates that it arose about 10,000 to 40,000 years ago, a timeframe that includes rapid expansion of human populations, travel, and the start of agriculture. Concluded the researchers, "Our data show that the creation of the 7R allele was an unusual, spontaneous mutation, which became an advantage for humans. Because it was an advantage, the gene became increasingly prevalent."

**SOURCE:** Yuan-Chun Ding et al. January 8, 2002. Evidence of positive selection acting at the human dopamine receptor D4 gene locus. *Proceedings of the National Academy of Sciences* 99(1):309-314.

**WORKSHEET:**

1. The 7R variant of the DRD4 dopamine receptor gene arose during the time of
    a. the Cro-Magnons
    b. *Australopithecus afarensis*
    c. Otzi the iceman
    d. The hominoids
    e. The Flintstones


2. How does selection for this trait differ from balanced polymorphism?

3. Explain how the 7R allele may have become more prevalent over time.




4. What might be some practical applications of screening for the 7R allele? What might be a danger of doing this?

# PART FIVE

# IMMUNITY AND CANCER

**CHAPTER 16**     **GENETICS OF IMMUNITY**
                               Five little piggies

**CHAPTER 17**     **GENETICS OF CANCER**
                               Adult T cell leukemia
                               Li-Fraumeni family cancer syndrome
                               Multiple endocrine neoplasia
                               Thyroid cancer

# FIVE LITTLE PIGGIES — CHAPTER 16

On Christmas day 2001, five very special piglets were born in Scotland, at the same facility where Dolly the famed cloned sheep was born five years earlier. The piglets were genetically modified to lack one copy of a gene that encodes an enzyme ($\alpha$-1,3-galactosyl transferase) that normally places a particular sugar on cell surfaces. As a result, half the normal amount of enzyme is made, and the pigs' cells have half the normal number of sugar molecules on their surfaces.

The sugar (galactose) is not on cell surfaces of the Old World primates, which includes humans. As a result, the human immune system recognizes this sugar as nonself, and launches a vigorous hyperacute rejection reaction against cells sporting this sugar. Specifically, the immune response attacks endothelial (lining) cells that form blood vessels that are part of a transplanted organ, turning them black in just minutes and cutting off the blood supply. Other molecules on pig cell surfaces can also induce hyperacute rejection. A few months before the Yuletide births in Scotland, U.S. researchers reported that they had cloned pigs that lacked one copy of this gene. (The Scottish research "knocked out" only the one gene; those piglets were not cloned.)

## SOURCES:

Butler, Declan. January 10, 2002. Xenotransplantation experts express caution over knockout piglets. *Nature* 415:103-104.

Kaiser, Jocelyn. January 4, 2002. Cloned pigs may help overcome rejection. *Science* 295:25-26.

## WORKSHEET:

1. How does the transplant of pig tissue to human differ from an allograft?




2. Said the research director of PPL Therapeutics, the Scottish company that created the five piglets, "The promise of xenotransplantation is now a reality." How did he overstate the importance of the pigs?

3. A further genetic manipulation that is still required to make either of the pigs described above better sources of transplant material is to:




4. Another potential risk of xenotransplantation, besides the hyperacute rejection response of humans, is _____
_____ .

# ACUTE T CELL LEUKEMIA   CHAPTER 17

The cancer cells of people who develop acute (rapid onset and progression) T cell leukemia have a chromosome 11 that is cut at a certain locus in the long arm and a piece placed on chromosome 14. The gene that is disrupted on chromosome 11 encodes a T cell receptor. In one affected boy, Raheem, the site on chromosome 14 where the chromosome 11 material inserts encodes a growth factor. In Maria, who also has this type of cancer, chromosome 11 exchanges parts with chromosome 8.

**SOURCE:** OMIM 151390

1. The type of chromosome aberration that Raheem has is a

   _____.

2. The type of chromosome aberration that Maria has is a

   _____.

3. Explain how Raheem's cancer arose.

4. Are these cancers likely due to activation of an oncogene, or deletion or inactivation of a tumor suppressor gene? How do you know?

# LI-FRAUMENI FAMILY CANCER SYNDROME         CHAPTER 17

About 100 families worldwide have the Li-Fraumeni family cancer syndrome. Affected people inherit a germline mutation in an autosomal gene called *p53*. Cancer develops when the second *p53* gene is mutated in somatic tissue. The p53 protein normally functions as a tumor suppressor.

In one Li-Fraumeni family, 25-year-old Clint has bone cancer. He had two sisters. One, Martha, died of breast cancer when her son David was 8 years old. Clint's other sister, Tina, died of osteosarcoma (bone cancer) at age 19. A brother, Nelson, is healthy. Their father died at age 27, also of bone cancer.

Geneticists sequence Clint's *p53* gene and find an insertion of one extra cytosine in a stretch of four cytosines. The result is a p53 protein shortened by 212 amino acids. Clint's young daughter Jill and nephew David, who are healthy, are tested for the mutation. They both have it. Nelson is tested and does not have it.

**SOURCE:** OMIM 151623

**WORKSHEET:**

1. What must happen for Jill or David to develop cancer?

2. How can the mutation in this family, involving only one DNA base, cause such a drastic change in the encoded protein?

3. Does the mutation that causes Li-Fraumeni family cancer syndrome overexpress the p53 gene or underexpress it?

4. "Whew, I can't get cancer," concluded Nelson with relief after his *p53* gene test. Is he correct? Why or why not?




5. Jill and David are children, yet their parents know that they face a very high risk of developing certain types of cancers. How can this information be used to protect their health?




6. How does the wild type version of the *p53* gene function?

# MULTIPLE ENDOCRINE NEOPLASIA — CHAPTER 17

Oprah Q. sees her family physician because of severe headaches, blurry vision, sudden sweating, and rapid heartbeat. Sometimes her blood pressure is elevated. The doctor orders blood and urine tests. A few days later, an X ray of Oprah's adrenal glands confirms what the doctor suspected – she has a small tumor in one adrenal gland, a condition called pheochromocytoma. The tumor is small, and after removal, the symptoms vanish.

Pheochromocytoma is often part of an inherited cancer syndrome called multiple endocrine neoplasia type 2 (MEN2). It affects glands of the endocrine system, particularly the adrenal glands and the thyroid. Another common manifestation is a precancerous condition of the parathyroid glands called parathyroid hyperplasia. The thyroid and parathyroid glands are located in the neck; the adrenal glands sit atop the kidneys. The tumors and growths associated with MEN2 have wide-ranging bodily effects because the hormones that they overproduce affect many organs.

Oprah did not realize that she had inherited MEN2, an autosomal dominant condition with a penetrance of 80%, because she didn't connect the various health problems in her family. Her brother Max had a thyroid tumor, but she does not speak to him often and did not know of his illness. Her sister Mindy had a "mild parathyroid problem" as a teenager, but no one had mentioned MEN2. Oprah, Max, and Mindy's father had died of cancer, but by the time it had been diagnosed, it had spread to his liver. The cause of death was officially liver cancer, but it could have begun in the adrenal or thyroid glands, both of which were also affected. The doctor, knowing that pheochromocytoma often affects several family members, takes a family history and concludes that the tumors and overgrowths are related, and part of MEN2.

The pedigree for this family is:

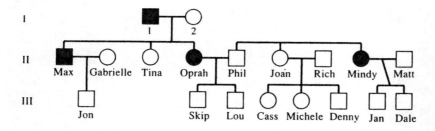

**SOURCE:** OMIM 171400

**WORKSHEET:**

1. Why is it important for Oprah to know that her cancer is not an isolated case, but part of an inherited syndrome?




2, In one study conducted in the 1980s, sixteen families were found to have a small deletion of chromosome 20p that segregates with members who have MEN2 but not with healthy members. The human genome project revealed a second causative gene on chromosome 10. These results mean that MEN2 is

        a. genetically heterogeneic
        b. pleiotropic
        c. incompletely penetrant
        d. autosomal dominant
        e. caused by a trisomy

3. The chance that Jan or Dale develops an endocrine tumor or overgrowth is

_____ .


4. Is a gene that causes MEN2 more likely to be an oncogene or a tumor suppressor? Cite a reason for your answer.

# THYROID CANCER

**CHAPTER 17**

Anaplastic thyroid carcinoma is a life-threatening form of thyroid cancer (most other forms are nearly 100% treatable). It is associated with very small deletions on various chromosomes. A research study detected the following deletions in five patients:

| Patient | Chromosomal Microdeletions |
| --- | --- |
| A | 17p, 19p |
| B | 11q, 17p, 19p |
| C | 22q, 19p, 9p, 11q, 17p |
| D | 17p |
| E | 11q, 17p, 22q, 9p, 19p, 1q |

## SOURCE:

Kitamura, Yutaka et al. March 2000. Allelotyping of anaplastic thyroid carcinoma: frequent allelic losses on 1q, 9p, 11, 17, 19p and 22q. *Genes, Chromosomes and Cancer* 2:244.

## WORKSHEET:

1. The order of chromosomal deletions that occurs as this type of cancer progresses is _____ .

2. Which person is probably the sickest? How do you know this?

3.. Some of these microdeletions are also present in other forms of cancer. Explain how this can happen.

4. Is a deletion more likely to be associated with a tumor suppressor gene or an oncogene? Cite a reason for your answer.

# PART SIX

# GENETIC TECHNOLOGY

**CHAPTER 18**     **GENE MODIFICATION**
                             Hemophilia A and B
                             Infertility drugs

**CHAPTER 19**     **GENE THERAPY AND GENETIC COUNSELING**
                             Gene doping
                             Hemophilia B
                             Newborn screening

**CHAPTER 20**     **AGRICULTURAL BIOTECHNOLOGY**
                             Designer potatoes
                             Transgenic tobacco

**CHAPTER 21**     **REPRODUCTIVE TECHNOLOGIES**
                             Charcot-Marie-Tooth disease
                             Male infertility

**CHAPTER 22**     **THE HUMAN GENOME PROJECT**
                         **AND GENOMICS**
                             Diffuse large B cell lymphoma
                             Muscle DNA microarray

# HEMOPHILIA A and B — CHAPTER 18

Hemophilia B was first described symptomatically in 1952 in a 16-year old male. Later that year, the underlying clotting malfunction was further described in a 5-year old boy named Christmas, and hence the condition became known as Christmas disease. It is a deficiency of clotting factor IX, which is encoded by a gene on the X chromosome that is very close to the gene for clotting factor VIII, which is responsible for the more common hemophilia A.

Tyshawn and Chaniqua meet at the hemophilia clinic of a regional medical center. Each is receiving a recombinant clotting factor, Tyshawn factor IX for hemophilia B, and Chaniqua factor VIII for hemophilia A. Chaniqua is concerned. She is taking a course in genetics, and learns that the clotting factor she receives comes from Chinese hamster ovary cells that have been genetically modified to produce the human clotting factor that she lacks. She is worried that she will have an allergic reaction because she sneezes from hamsters, and that she could contract a viral infection from the source of the drug. Tyshawn assures her that the recombinant drug is actually safer than taking clotting factor pooled from donors, as was done before 1984.

After many months, Tyshawn and Chaniqua start to talk about marriage, although they are each concerned that a child of theirs could have either or both of their clotting disorders.

**SOURCES:** OMIM 306700 (hemophilia A)
OMIM 306900 (hemophilia B)
OMIM 134510 (factor VIII and IX combined deficiency)
http://www.wyeth.com/products/benefix.asp

## WORKSHEET:

1. Two reasons that a recombinant protein-based drug is safer than the same drug obtained from non-human animals or from pooled human donors are

_____ and

_____.


2. Another young person at the clinic, Enrico, has factor VIII and IX combined deficiency. Suggest a way that this condition might arise.

3. What would happen if Tyshawn was mistakenly given recombinant factor VIII, or Chaniqua given recombinant factor IX?




4. Can you determine which of Tyshawn and Chaniqua's parents has a clotting disorder or is a carrier for one? If so, identify them.




5. If Tyshawn and Chaniqua have a son, the probability that he has factor VIII deficiency is _____ because _____. The chance that he inherits factor IX deficiency is _____ because _____ _____.


6. After meeting with a genetic counselor, the couple decides that when they want to have children, they may use a sperm-sorting procedure that will greatly increase the chances of them conceiving a female. Why do you think they would do this?

# INFERTILITY DRUGS — CHAPTER 18

A biotechnology company offers the following products to regulate a woman's menstrual cycle so that she will be able to have oocytes retrieved for use in *in vitro* fertilization (see Chapter 21):

- recombinant human follicle stimulating hormone (r-hFSH)
- recombinant human luteinizing hormone (r-hLH)
- recombinant human chorionic gonadotropin (r-hCG)

A physician prescribed r-hFSH and r-hLH to cause one woman to superovulate (produce more than one mature oocyte), and r-hFSH with r-hCG to stimulate ovulation in a woman who does not often ovulate. The human proteins are produced in Chinese hamster ovary cells growing in culture.

**SOURCE:** http://www.serono.com

**WORKSHEET:**

1. Before the advent of recombinant DNA technology, these proteins were obtained from the urine of female donors. How is the recombinant method safer?

2. If these drugs are human versions of these fertility hormones, why are they considered to be recombinant?

3. What is a potential complication of using similar hormones from non-human animals?

4. What must these three hormones have in common to be able to be manufactured using recombinant DNA technology, compared to a steroid hormone such as estrogen or progesterone?

# GENE DOPING CHAPTER 19

First there was blood doping. An elite athlete would store blood months before competition, then receive a transfusion shortly before the event. The extra red blood cells boost aerobic capacity, and therefore can enhance athletic performance. Next came illicit use of the hormone erythropoietin (EPO), made using recombinant DNA technology. The kidneys normally secrete EPO, which stimulates the bone marrow to produce more red blood cells. Several cyclists died after taking too much recombinant EPO. The drug thickened their blood and increased the risk of stroke and other inappropriate, dangerous blood clotting. In experiments, baboons given EPO produce so many extra red blood cells that their blood had to be periodically diluted! However, EPO enjoys such a prominent reputation as a performance enhancing agent that Olympic officials fear that athletes may seek a gene therapy so that their bodies will produce extra EPO.

**SOURCE:** David Adam, December 6, 2001. Gene therapy may be up to speed for cheats at 2008 Olympics. *Nature* 414:569-570.

**WORKSHEET:**

1. Devise a protocol to deliver somatic gene therapy so that an athlete can produce EPO.

2. Do you think that "gene doping" is a valid use of gene therapy?

3. Two possible dangers of using gene doping to produce EPO are:

4. What technique would officials have to use to detect gene doping?

# HEMOPHILIA B            CHAPTER 19

A gene therapy trial to treat the blood clotting disorder hemophilia B was suspended when researchers discovered the virus used to transfer the healing gene, adeno-associated virus (AAV), in the semen of volunteers. The gene therapy consisted of the clotting factor IX gene delivered in the viral vector to the livers of men with severe hemophilia. The virus was detectable in semen for 10 weeks following the treatment. Semen consists of many sperm cells, some white blood cells, and secretions. AAV was detected in the semen, but not in sperm cells.

**SOURCES:** OMIM 306900

Neil Boyce. December 13, 2001. Trial halted after gene shows up in semen. *Nature*, 414:677.

Eliot Marshall. December 14, 2001. Panel reviews risks of germline changes. *Science* 294:2268-2269.

**WORKSHEET:**

1. AAV is one of the safest vectors for gene therapy. What, then, is the likely reason for the concern over this gene therapy trial?

2. An alternative way to treat a clotting disorder that uses biotechnology is

_____.

3. Do you think that it would be ethical to restrict use of this gene therapy to women, or to men who are infertile or who have had a vasectomy?

4. If the transgene shows up in semen and actually enters sperm cells, do you think that the gene therapy is still feasible, if the man uses a condom or does not have sex until the transgene and AAV are no longer present in his sperm? About how long would this take? (see Chapter 3).

5. How might the AAV and its genetic cargo have found its way to white blood cells in semen?

# NEWBORN SCREENING — CHAPTER 19

In the early 1960s, State University of New York at Buffalo microbiologist Robert Guthrie developed a simple blood test performed on newborns to detect phenylketonuria (PKU), an inborn error of metabolism that can be treated with a successful diet to prevent mental retardation if it is started as soon after birth as possible. The diet is extremely low in the amino acid that builds up in the condition, causing the symptoms. Guthrie's son had the disorder. The Guthrie test quickly became mandatory, and over the years, other tests were performed on the same blood sample, taken from the newborn's heel. Now, a technique called tandem mass spectrometry is being used to screen for 43 inborn errors of metabolism. The technique analyzes blood using two mass spectrometers, a tool commonly used in analytical chemistry. The first "mass spec" catalogs general types of molecules in the sample, and the second identifies 65 specific products of metabolism that provide clues to inherited illnesses. The test takes minutes and costs only a few dollars per sample. Implementing these tests, however, requires appropriate follow-up treatments and genetic counseling services. Some states that have the equipment have yet to institute screening because they lack the infrastructure to follow the families of newborns diagnosed with disease.

Nora Waananen died at 4 months of age, of a rare inborn error called long-chain 3-hydroxyacyl-CoA dehydrogenase deficiency (LCHAD). One day the northern California infant seemed unusually sleepy, and went into a coma and died within a day. She was diagnosed following an autopsy, and because the mode of inheritance is autosomal recessive, the parents deduced that they must be carriers. The mother, Sirpa, investigated and found that California indeed had the two mass specs, but the machines were not being used because programs for follow-up had never been set up. She is leading a national effort to make tandem mass spectrometry for newborn screening mandatory in all states, pointing out that it is unethical for a newborn with a certain illness to be treated and survive in Massachusetts, but for this not to be the case for a baby in Maine with the same disorder. Five northeastern states are currently running pilot programs in which parents must "opt out" and specifically request that their newborns not be tested. So far, only 3 percent choose this option. In addition, parents are informed that an inborn error of metabolism is present only if there is a way to treat the illness.

**SOURCES:** OMIM 261600, 201450

Eliot Marshall. December 14, 2001. Fast technology drives new world of newborn screening. *Science* 294:2272-2274.

**WORKSHEET:**

1. A problem with not informing parents of a disorder in their newborn that isn't treatable is:

2. What criteria do you think should be considered in deciding which disorders to screen for, especially since the human genome project will yield information that can be used to develop many new tests?

3. When the government of Iceland planned a national health database in which citizens would have to "opt out," there was a bioethical outcry that this robs people of choice. Do you think that an "opt out" policy makes sense for newborn screening? Cite a reason for your answer.

# DESIGNER POTATOES    CHAPTER 20

Everyone loves potatoes, in one form or another. Not all potatoes, however, are equally pleasing to the human palate. Some are coarse and tasteless, others contain natural toxins, many are susceptible to a variety of pests, and some species have very small tubers.

## WORKSHEET:

1. Compare how *Ti* plasmid mediated transgenesis, protoplast fusion, and mutant selection could be used to develop a strain of edible potato that resists a specific viral infection.




2. A gene from a wild variety of potato enables the plant to resist damage by leaf hoppers, potato tuber moths, and the Colorado potato beetle. The characteristic is called "glandular trichome trait," and it causes the plant to become covered with specialized epidermal outgrowths that deter these pests. Describe the steps that you would use to transfer this trait to an edible strain of potato.




3. A trait called "high dry matter" is valuable in potatoes that are to be dried out and used to make chips, but it naturally exists in a strain of potato that has a bitter taste. Explain a major difference in trying to develop a high dry matter good-tasting potato using traditional agriculture, versus using a biotechnology such as transgenesis.

# TRANSGENIC TOBACCO  CHAPTER 20

Tobacco is often used as a model organism in plant biotechnology. That is, experiments are conducted to see if a particular manipulation works, not necessarily to change smoking tobacco.

In one experiment, tobacco leaf cells growing in culture are transformed with a *Ti* plasmid that contains:

- a promoter from a tomato gene
- a gene from a cyanobacterium (a simple one-celled prokaryote that photosynthesizes) that produces an enzyme that greatly increases photosynthetic efficiency
- an antibiotic resistance gene
- genes that encode "transit peptides" that transfer a protein into a chloroplast, where photosynthesis occurs

The resulting plants grow unusually large leaves faster than normal.

In a second set of experiments done by a different team, tobacco leaf cells are transformed with a *Ti* plasmid that contains a bacterial gene, called *nfsl*, that encodes an enzyme called nitroreductase. The enzyme enables the plant to detoxify trinitrotoluene, which is an organic compound better known by its initials, TNT, as an explosive. TNT contaminates many sites near munitions factories. Some plants can naturally break down TNT and other toxins, thanks enzymes that enable them to grow in the presence of similar toxins released by soil-dwelling microbes. The researchers thought that creating transgenic tobacco would greatly increase the plant's detoxifying capability. This is indeed the case. Seeds of transgenic plants grown in liquid medium containing concentrations of TNT that would kill normal seeds not only survive, but they germinate into healthy plants, as the level of TNT in the medium greatly diminishes.

## SOURCES:

Hannink, Nerissa et al. December 2001. Phytodetoxification of TNT by transgenic plants expressing a bacterial nitroreductase. *Nature Biotechnology* 19:1168-1172.

Meagher, Richard B. November 2001. Pink water, green plants, and pink elephants. *Nature Biotechnology* 19:1120-1121.

Miyagawa, Yoshiko, et al. October 2001. Overexpression of a cyanobacterial fructose-1,6-sedoheptulose-1,7-bisphosphatase in tobacco enhances photosynthesis and growth. *Nature Biotechnology* 19:956-969.

**WORKSHEET:**

1. What is the function of each component of the *Ti* plasmid used to boost the photosynthetic capability of tobacco?

2. What types of tests must be done to follow up on the safety and efficacy of the two types of manipulations described here?

3. TNT consists of an organic ring to which three nitrate (a nitrogen bound to two oxygens) groups attach. In some polluted lagoons and streams, one nitrate group is removed, perhaps by microbial enzymes, to produce dinitrotoluene. This "DNT" turns the water pink. In the second experiment, the water in which the transgenic tobacco is immersed is not pink, but white. What do you think has happened?

# CHARCOT-MARIE-TOOTH DISEASE    CHAPTER 21

Charcot-Marie-Tooth disease produces a progressive loss of sensation in the distal parts of the arms and legs, beginning in adulthood. The disorder is usually not as severe as other neuromuscular conditions. It is usually inherited as an autosomal dominant trait. Although at least 8 different genes can cause it, nearly half of all cases are caused by a duplication of 1500 kilobases on chromosome 17.

In one experiment, preimplantation genetic diagnosis was performed on an 8-celled blastocyst. One cell was removed, and the polymerase chain reaction (PCR) used to look for the telltale duplication. Because PCR did not detect this mutation, the remaining 7-celled blastocyst was implanted into the woman who donated it, and continued developing. Nine months later, Kenneth was born, apparently healthy.

**SOURCES:** OMIM 118200

Kim, Hak-Soon. November 1999. Preimplantation diagnosis of the B1 integrin knockout mutation as a model for aneuploid gene testing. *Human Genetics*, vol. 105, p. 480.

**WORKSHEET:**

1. Is Kenneth expected to have inherited Charcot-Marie-Tooth disease? Why or why not?

2. Explain how PCR and Southern blotting would be used to detect the specific mutation that causes the most common form of this disorder in a blastocyst.

3. Do you think that preimplantation genetic diagnosis should be reserved for lethal or very serious medical conditions with early onset, or also be used for milder conditions, such as Charcot-Marie-Tooth disease?

4. How might a meiotic error cause Charcot-Marie-Tooth disease to appear in a family where it hasn't occurred before?

# MALE INFERTILITY                                              CHAPTER 21

A study to assess sperm quality investigated the sperm cells of 765 infertile men and 696 fertile men, with the following results:

| Sperm quality | Infertile Men | Fertile Men |
| --- | --- | --- |
| Concentration | <13.5 million/milliliter | > 48.0 million/milliliter |
| Motile cells | < 32 % | > 63 % |
| Normal morphology | < 9 % | > 12 % |

Three couples meet at a fertility clinic. The women are fertile. Harold Resnick has a sperm count of 21.4 million cells per milliliter, with 28% motile and 95% with abnormal morphology (shape). Peter Fromm has a sperm count of 8.6 million per milliliter, but 60% of the cells are of normal morphology and are motile (although these may not be the same 60%). Ralph Baron learns that he has azoospermia. This means that he has no mature sperm cells, but he does have spermatids.

**SOURCE:** David S. Guzick et al. November 8, 2001. Sperm morphology, motility, and concentration in fertile and infertile men. *The New England Journal of Medicine* 345(19):388-393.

## WORKSHEET:

1. Intracytoplasmic sperm injection (ICSI) could help _____.

2. Who is more likely to become a father sooner, Harold Resnick or Peter Fromm? Cite a reason for your answer.

3. The researchers qualify their results by stating that they are not absolute enough to be used to diagnose infertility, because of the overlaps and gaps. What do you think they meant?

# DIFFUSE LARGE B CELL LYMPHOMA — CHAPTER 22

Chemotherapy can cure fewer than half of patients with diffuse large B cell lymphoma (DLBCL), a cancer of the white blood cells also known as non-Hodgkin's lymphoma. Currently, physicians determine how high the doses should be based on a standard profile that considers age, stage of the cancer, the number of sites in the body that contain cancerous B cells, and other factors. The cells of all patients with DLBCL look alike when stained. Using this method of classifying patients results in giving very high doses of chemotherapy to many patients whose cancers do not respond.

Genetics can tailor the subtyping of cancers. Researchers at the Whitehead Institute are pioneering genetic and genomic approaches to better diagnosing cancer. For DLBCL, this includes screening for variants of 13 genes known to affect B cells – such as genes that encode signaling molecules, cellular adhesion molecules, cell cycle proteins such as kinases and cyclins, oncogenes, tumor suppressors, and genes that control apoptosis (programmed cell death) and angiogenesis (ability to build a blood supply). In addition to screening for variants of these 13 genes, the researchers developed DNA microarrays to test for the expression of 6,817 genes. If a gene is expressed in a cancerous B cell but not in a healthy B cell, then it can be used to refine diagnosis of this type of cancer.

By conducting these tests on 77 people diagnosed with DLBCL by standard criteria, the researchers found that the subjects fell into two broad categories. In one group, the cancerous B cells had gene expression profiles that were very similar to those of normal B cells found in lymphoid organs. These people had a 70 percent 5-year survival rate, and tended to respond well to standard chemotherapy. The other group had a 12 percent 5-year survival rate, and did not do well on chemotherapy.

**SOURCES:** OMIM 109565

Margaret A. Shipp et al. January 2002. Diffuse large B-cell lymphoma outcome prediction by gene-expression profiling and supervised machine learning. *Nature Medicine* 8(1):68-74.

Laura J. Van't Veer and Daphne De Jong. January 2002. The microarray way to tailored cancer treatment. *Nature Medicine* 8(1):13-14.

**WORKSHEET:**

1. An editorial accompanying the report on genetic testing for DLBCL called the microarray approach "a prototype for cancer management in the future." Why is the genomic approach likely to lead to more useful subtyping for this type of cancer than current methods?

2. If all cells contain the same complement of genes, how can a microarray test detect differences between cancerous and noncancerous cells from the same individual?

3. Some patients with DLBCL have a translocation between chromosomes 14 and 18 that leads to overexpression of a gene, *bcl2*, that normally blocks apoptosis. How does this lead to cancer?

4. How can the DNA microarray test identify patients who overexpress the *bcl2* gene but do not have the translocation?

5. Propose a study to determine whether the genetic and genomic tests to subtype this cancer actually help patients.

# MUSCLE DNA MICROARRAY — CHAPTER 22

The human genome project revealed a minimal set of 400 genes whose products are essential for the functioning of a muscle cell. DNA microarrays to probe muscle function include some of these genes, which encode:

- myoglobin, which carries oxygen in muscle tissue
- actin and myosin, which form the sliding filaments that provide contractility
- dystrophin, which is the protein missing in Duchenne muscular dystrophy, which maintains the structural integrity of the muscle cell membrane
- dystrophin-associated proteins, which bind dystrophin
- troponin and tropomyosin, which associate with actin and myosin and make their contractility possible

The microarray also includes gene variants known to cause certain neuromuscular disorders such as myotonic dystrophy and inherited forms of Parkinson disease, and gene variants for cardiovascular function, including alleles of several genes that contribute to hypertension. The muscle microarray for women includes mitochondrially encoded proteins that control energy level and utilization. Future versions of the muscle microarray will include a series of genetic markers identified among Olympic athletes.

**SOURCE:** Bortoluzzi, Stefania et al. March 2000. The human adult skeletal muscle transcriptional profile reconstructed by a novel computational approach. *Genome Research*, vol. 10, p. 1.

## WORKSHEET:

1. An athletic young couple wishes to start a family, but want to be assured that their child will be physically able to join them in swimming, skiing, running, hiking, biking, hang gliding, and mountain climbing. Do you think that it is wise to use this DNA microarray to help them "select" a child with the desired characteristics? Cite reasons for your answer.

2. A healthy young woman undergoing the muscle microarray test is astonished to learn that she is a carrier of Duchenne muscular dystrophy. She does not have any affected relatives. Explain two ways to account for the fact that she is the only one in her family known to carry this gene.

3. The chance that a son of this woman will inherit Duchenne muscular dystrophy is _____.

4. A muscle microarray for a man would not include mitochondrial genes because _____ .

5. Do you think that the future test based on genetic markers from Olympic athletes will provide meaningful information? Why or why not?

6. Actin is a major constituent of muscle, but is also part of the cytoskeleton of all cell types. How can DNA microarrays distinguish muscle cells from other cell types?

7. A young couple learns that each of them is a carrier of a condition in which the same dystrophin-associated protein is abnormal. The genetic counselor who interprets the muscle DNA microarray tells them that each of their children faces a 1 in 4 chance of inheriting a rare form of muscular dystrophy. They are confused – isn't muscular dystrophy X-linked? Explain how the rare form of the illness is inherited.

# PART SEVEN

# CONNECTIONS AND SYNTHESIS

Argininemia

Blepharophimosis, ptosis, epicanthus inversus syndrome (BPES)

Complement component 2 deficiency

Dilated cardiomyopathy

Ehlers-Danlos syndrome type IV

Familial mental retardation (ATR-16)

Hereditary multiple exostoses

Lysinuric protein intolerance

Muscular dystrophy

Nephrolithiasis

Prenatal microarray screen

Pseudohermaphroditism

Silver-Russell syndrome

Smith-Lemli-Opitz syndrome

Tangier disease

Townes-Brocks syndrome

Venous thrombosis

## ARGININEMIA

### KEY WORDS

Enzyme
Exon
Genetic code
Genetic manipulation

Inborn error of metabolism
Mode of inheritance
Multiple alleles
Mutation

Japanese researchers examined mutant alleles of a gene that encodes an enzyme, liver-type arginase, in four patients. The normal protein catalyzes the breakdown of arginine, an amino acid. In argininemia, lack of the enzyme causes progressive mental retardation, spastic limb movements, seizures, and growth retardation.

A cDNA revealed that the coding portion of the gene specifies 322 amino acids. The entire gene is 11.5 kilobases and is on chromosome 6q. Argininemia affects both sexes and is inherited from carrier parents.

Patient A is homozygous for a G mutated to an A at DNA base 365 in the liver-type arginase gene. Patient B is homozygous for a G to C mutation at base 703, which substitutes one amino acid for another. Patient C has patient A's mutation and patient B's mutation. Patient D has patient A's mutation in one allele, and the other allele is a deletion of a C at position 842.

The researchers evaluated the phenotype associated with each allele by expressing it in *E. coli*. Patient A's abnormal protein is too short. The other mutations yield proteins of normal length that are unstable or otherwise nonfunctional.

**SOURCE:** OMIM 207800

**WORKSHEET:**

1. The mode of inheritance of argininemia is _____.

2. The patients who are heterozygotes for the argininemia gene are _____ and _____.

3. The patients who have missense mutations are _____ and _____.

4. Patient A's liver-type arginase is too short because _____

_____.

5. Human liver-type arginase can be synthesized and expressed in *E. coli* because
    a. *E. coli* have livers
    b. The genetic code is universal
    c. The disorder is recessive
    d. *E. coli* also uses arginine
    e. The genetic code is triplet

6. The argininemia gene has enough bases beyond those in exons to encode

   _____ more amino acids.

## BLEPHAROPHIMOSIS, PTOSIS, EPICANTHUS INVERSUS SYNDROME (BPES)

### KEY WORDS

| | |
|---|---|
| Deletion | Mode of inheritance |
| DNA microarrays | Mutation |
| Karyotype | Sex limited trait |
| Linkage | Translocation |

BPES causes low-set ears, arched brows, droopy overgrown eyelids and premature ovarian failure. In 26 2-generation pedigrees analyzed in one study, 86 females and 92 males are affected among the offspring, and in each pedigree, one parent is affected.

The gene variant that causes BPES is almost identical to a gene that causes male goats to be hornless, and female goats to develop male genitals. The goat gene is called PIS, for polled intersex syndrome. The genes that surround the BPES gene in humans, and those that surround the corresponding gene in goats, are also the same.

**SOURCES:** OMIM 110100

Lewis, Ricki. April 3, 2000. Of hornless goats and droopy eyelids. *The Scientist* 14(7):2.

Schibler, Laurent et al. March 2000. Fine mapping suggests that the goat polled intersex syndrome and the human blepharophimosis ptosis epicanthus syndrome map to a 100 kb homologous region. *Genome Research* 4(7):311.

**WORKSHEET:**

1. A likely mode of inheritance for BPES is _____.

2. Explain the evolutionary significance of the correspondence between the human BPES gene and the goat PIS gene. (Use your imagination; even researchers don't know the answer!)

3. Some families with BPES have a reciprocal translocation that disrupts a gene in chromosome 3q. Sporadic cases are often associated with a small deletion in chromosome 3q. Most affected families, however, have normal karyotypes. How can all of these families have the same syndrome?

4. BPES is nearly always transmitted from the father. Two hypotheses can explain this inheritance pattern: the disease is sex limited, or it is caused by deletion of two tightly linked genes, one conferring the facial characteristics, the other conferring female infertility. Does the information that goats have a similar syndrome enable a distinction to be made between these two hypotheses? Cite a reason for your answer.

# COMPLEMENT COMPONENT 2 DEFICIENCY

## KEY WORDS

Control of gene expression
Mode of inheritance

Multifactorial inheritance
Pedigree analysis

The C2 gene encodes the C2 protein, which is part of the complement system, a component of the immune response. The C2 gene resides in the HLA (human leukocyte antigen) gene complex on chromosome 6.

A person with two mutant C2 alleles suffers frequent bacterial infections. Carriers can be detected by determining the HLA type. Two types of defects in C2 gene function are known. In type I C2 deficiency, found in 94 % of affected families, no C2 protein is detectable. In type II C2 deficiency, cells cannot secrete C2 protein.

C2 deficiency can be difficult to diagnose because many conditions can impair immunity against bacterial infection. In the Fonebone family, the frequent severe infections in two brothers led to the diagnosis. Jerome Fonebone was hospitalized six times in three years for strep throat, cellulitis, croup, and various unexplained fevers. His brother Antone had repeated bouts of pneumonia and sinusitis. A sister, Mona, is very healthy, as is her mother Joan. Their father, Tyrone, however, suffers often from bronchitis, sinusitis, and pneumonia. HLA typing revealed that Joan is a carrier for C2 deficiency.

**SOURCE:** OMIM 217000

**WORKSHEET:**

1. The mode of inheritance for C2 deficiency is _____.

2. Draw a pedigree for the Fonebone family.

3. Is Mona a carrier of C2 deficiency?

4. In what sense is C2 deficiency a multifactorial trait?

5. How could you distinguish between types I and II C2 deficiency at the molecular level?

6. How could C2 deficiency be non-penetrant?

# DILATED CARDIOMYOPATHY

## KEY WORDS

cDNA
Chromosome
Genetic heterogeneity

Mutation
Pleiotropy
Variable expressivity

Dilated cardiomyopathy is a general term for several disorders. About 35% of affected individuals inherit the condition, which may be autosomal dominant, autosomal recessive, or X-linked recessive. The condition can cause congestive heart failure, blood clots that obstruct circulation, heart failure, arrhythmias, and, in 28% of cases, sudden death. Affected individuals can have any combination of symptoms. Mutations in at least 5 genes can cause this condition. One such gene is on chromosome 1q, and may cause dilated cardiomyopathy when it has any of the following missense mutations:

> arginine to glycine
> asparagine to lysine
> glutamate to glycine

**SOURCES:** OMIM 1152001, 600884, 601154, 602067, 300069

Fatkin, Diane et al. December 2, 1999. Missense mutations in the rod domain of the laminin A/C gene as causes of dilated cardiomyopathy and conduction-system disease. *The New England Journal of Medicine* 341:1715.

## WORKSHEET:

1. What is the evidence that dilated cardiomyopathy is genetically heterogeneic?

2. Three specific mutations that could account for the form of dilated cardiomyopathy that is associated with the gene on chromosome 1q are

_____, _____, and _____.

3. Why would it be difficult to identify whether a person has inherited dilated cardiomyopathy?




4. Is the gene on chromosome 1 that causes dilated cardiomyopathy on the long or short arm?




5. How would you prepare a cDNA to a gene that causes dilated cardiomyopathy?

# EHLERS-DANLOS SYNDROME TYPE IV

## KEY WORDS

Mode of inheritance
Mutation
Pedigree

Pleiotropy
Variable expressivity

Ehlers-Danlos syndrome refers to several disorders of collagen, the connective tissue protein that comprises much of the human body. In Ehlers-Danlos syndrome type IV, also called the "vascular type," various structures can suddenly rupture – most commonly arteries, intestines, and uteruses. Average age at death is 48 years. Other symptoms include easy bruising, thin skin, and distinctive facial features. The mutation that causes this condition is in a gene called *COL3A1*. It is autosomal dominant, and is variably expressive.

Collagen is normally a triple helix, cut from a longer molecule called procollagen. Overall collagen consists of many repeats of the amino acid sequences "glycine-X-Y," where X is usually proline and Y is a form of proline with a hydroxyl (OH) group added. These amino acids are small, which enables the tight triple helix to form. In people with Ehlers-Danlos syndrome type IV, some procollagen single helices are abnormal and cannot associate with normal ones. As a result, only an eighth of the collagen triple helices are normal, and this deficit causes the symptoms. The specific pattern of symptoms arises from where in the body the abnormal collagen triple helices form.

In the first generation of the Maguire family, Bruce died at age 53 of a burst aorta. He and his wife Martha had four children. Sheila bruises easily, as does her son Phil. Sheila's brother Dan is healthy. Her sister Shelley died from a ruptured uterus giving birth to David, and her brother Eric died at age 27 of a ruptured large intestine. A workup at a genetics clinic following Shelley's death led to diagnosis of Ehlers-Danlos syndrome type IV. Other family members can now be tested for blood vessel weakenings.

**SOURCES:** OMIM 225360

Pepin, Melanie et al. March 9, 2000. Clinical and genetic features of Ehlers-Danlos syndrome, type IV, the vascular type. *The New England Journal of Medicine* 342:673.

## WORKSHEET:

1. Draw a pedigree of the Maguire family. Identify unnamed individuals with numbers.

2. Which family members should be screened for the tendency to develop aneurysms (burst blood vessels)?

3. Many of the mutations that cause this condition replace glycine with a larger amino acid. How does this disrupt the collagen structure?

4. Phil is aware of the rupture problem in the family, but thinks that because his symptom is only bruising, that he need not be concerned about his future health, or about passing on the gene to offspring. Is this wise? Cite a reason for your answer.

5. The evidence that Ehlers-Danlos syndrome type IV is pleiotropic is that

_____.

# FAMILIAL MENTAL RETARDATION (ATR-16)

## KEYWORDS

Anticipation
FISH
Haplotypes

Linkage
Prenatal diagnosis
Translocation

A large family studied at a medical genetics center in Munich had 10 members with mental impairment. The four affected individuals in the fifth (most recent) generation were moderately mentally retarded, and the six affected relatives in the fourth generation were mildly retarded. The mental impairment affects male and females, and is passed from males to sons and daughters as well as from females to sons and daughters. The affected individuals in the fourth and fifth generations had similar facial features, including widely spaced eyes and broad noses.

    The combination of mental retardation and similar facial features prompted researchers to conduct a series of tests to link a syndrome to a particular part of the genome. Conventional chromosome analysis revealed an apparently normal karyotype, with no fragile sites. Next, the investigators conducted a whole genome linkage analysis using microsatellite repeats found on all 24 chromosome types. Microsatellites corresponding to chromosome 16 segregated with the affected family members. Additional chromosome 16 markers were added to the testing, narrowing down a candidate region to 16p. Use of several other markers, combined with FISH, finally revealed that the family had a balanced translocation in which the tips of chromosomes 16q and 3q exchanged parts. The abnormality was too small to be detected in a standard karyotype. Some of the individuals who inherited one translocated chromosome are mentally normal.

    Haplotypes for part of the extensive pedigree are:

**SOURCES:** OMIM 301040

Holinski-Feder, Elke et al. January 2000. Familial mental retardation syndrome ATR-16 due to an inherited cryptic subtelomeric translocation t(3; 16)(q29:p13.3). *The American Journal of Human Genetics* 66:16.

**WORKSHEET:**

1. Which arms of each chromosome are implicated in the translocation in this family?

2. Why does individual II1 have symptoms, but his sister does not?

3. What is the evidence that the mental retardation in this family is not due to trisomy 21 or fragile X syndrome?

4. A genetic mechanism that might account for the worsening of symptoms with each generation is _____.

5. Describe how a prenatal diagnosis technique might be used in this family to identify fetuses that have inherited the translocation. Why might this approach be limited in predicting mental retardation?

# HEREDITARY MULTIPLE EXOSTOSES

## KEY WORDS

Chromosome aberrations
Expressivity
Genetic heterogeneity

Mutation
Penetrance

Tina was diagnosed at age three with hereditary multiple exostoses, in which excess bone, capped with cartilage, forms at the ends of the long bones. Her arms and legs appeared deformed, and she had difficulty running, but was otherwise healthy. When Tina was ten, her brother, Evan, then four, was also diagnosed with the condition. Their parents did not appear to be affected. Tina and Evan have a very small deletion on the long arm of chromosome 8. Affected individuals in another family with this disorder have a translocation between chromosome 8 and chromosome 11. Yet another affected family has a nonsense mutation in the causative gene, and in this family, it is inherited in an autosomal dominant manner.

     The incidence of hereditary multiple exostoses varies in different populations, ranging from 1 in 1,000 among the Chamorros group on the island of Guam, to 14 out of every million births in the United Kingdom. The condition is also seen in horses, cattle, dogs, cats, lions, lizards, and even fossils of dinosaurs. The lesions continue to grow until puberty, and then they may shrink. The abnormality causes short stature and widening of the bones. Expression is very variable. Some children are so mildly affected that the disorder appears only as extra bony growths on X-rays – it does not interfere with activities.

**SOURCES:** OMIM 133700

Xu, Lei et al. July 8, 1999. Mutation analysis of hereditary multiple exostoses in the Chinese. *Human Genetics* 105:45.

**WORKSHEET:**

1. A third of cases of hereditary multiple exostoses are new mutations. How do we know that this is not the case in Tina and Evan's family?


2. A test that could have been performed on the parents to determine if Tina's condition is inherited or the result of a new mutation at the time of her diagnosis

would have been a _____.

3. This disorder might be unusually prevalent on Guam because:




4. Hereditary multiple exostoses is incompletely penetrant and variably expressive. This means that:

   a. females are more severely affected than males, and have excess bony material in more parts of their bodies.

   b. everyone who inherits the mutant allele has symptoms, although some individuals may be more severely affected than others.

   c. not everyone who inherits the mutant allele has symptoms, and of those that do, the degree of severity can vary.

   d. different types of mutations cause the disorder.

   e. it is recessive in one sex but dominant in the other.


5. Is hereditary multiple exostoses caused by a "loss of function" or a "gain of function" mutation? How do you know this?

# LYSINURIC PROTEIN INTOLERANCE

**KEY WORDS**

Consanguinity  
Founder effect  
Introns  
Mode of inheritance

A person with lysinuric protein intolerance cannot digest three of the twenty types of dietary amino acids, due to a defect in the small intestinal wall. Symptoms include poor weight gain, vomiting, diarrhea, enlarged liver and spleen, and coma. It is deadly in childhood. The condition is extremely rare, affects males and females, and skips generations. It is more common in Italy and Finland than in other countries due to founder effects. Consanguinity is present in some affected families.

**SOURCES:** OMIM 222700

Sperandeo, Maria Pia et al. January 2000. Structure of the SLC7A7 gene and mutational analysis of patients affected by lysinuric protein intolerance. *The American Journal of Human Genetics* 66: 92.

**WORKSHEET:**

1. Explain how a founder effect would make a rare disorder more common in some populations than others.


2. What is consanguinity, and why would it be more common among families with this disease than among the general population?


3. The likely mode of inheritance for this condition is _____.

4. Suggest a way to try to treat lysinuric protein intolerance.


5. A mutation involving an intron/exon splice site could cause this condition by

_____.

# MUSCULAR DYSTROPHY

## KEY WORDS

Haplotype
Mode of inheritance
Pedigree

Hilda and Roy are carriers of a form of muscular dystrophy whose causative gene maps to chromosome 1, which they learned from a genetic work-up of their affected daughter, Kelly. Their other two children, Dylan and Donna, are carriers, but haplotype analysis reveals that they do not have the same DNA sequence for this region of the chromosome. The haplotypes and pedigree are below. (Shading indicates a chromosome with the mutant allele.)

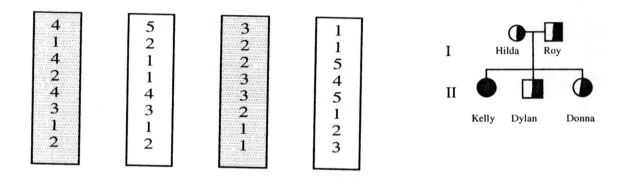

**SOURCE:** Brockington, Martin et al. February 2000. Assignment of a form of congenital muscular dystrophy with secondary merosin deficiency to chromosome 1q42. *The American Journal of Human Genetics* 66:428.

## WORKSHEET:

1. Draw possible haplotypes for Kelly, Dylan and Donna.

2. Write the haplotype for a child of Hilda and Roy who is not affected or a carrier.

3. The mode of inheritance is _____.

4. How would a haplotype of [ 4 1 4 | 1 4 3 1 2 ] arise in this family?

# NEPHROLITHIASIS

## KEY WORDS

Conditional probability
Consanguinity
Degree of relationship

Mode of inheritance
Pedigree analysis

Over a two year period, doctors at a medical center examined four young men with kidney stones and protein in their urine. Ned, Ted, Fred, and Jed each saw a different doctor. At a clinical staff meeting, the doctors realized that these patients had very similar symptoms. Each had been diagnosed differently probably because kidney stones are part of many syndromes. On further analysis, they were all diagnosed with nephrolithiasis, an inherited disorder.

The four men all had different last names, so it had not occurred to anyone that they might be related. Ned, Ted, and Fred's mothers, however, are sisters. These sisters, Kelly, Nelly, and Ellie, also have identical twin brothers, Sam and Cam who, like their sisters, are healthy. Ned, Ted, and Fred's maternal grandfather Red also had progressive kidney failure. Jed, the other young man with kidney stones, has a maternal grandmother who is a sister of Red. Jed's uncle Ed also has the kidney ailment. The family pedigree is:

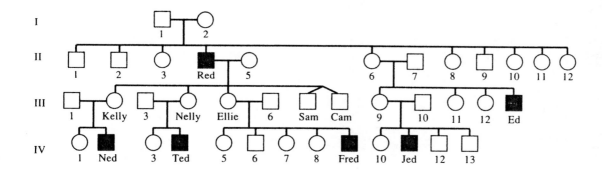

**SOURCE:** OMIM 310468

## WORKSHEET:

1. Indicate on the pedigree who must be carriers, if the condition is not dominant.

2. The most likely mode of inheritance is _____

because _____ .


3. Treatment for nephrolithiasis is a kidney transplant. Which individuals would be the most successful donors for Ned, Ted, Fred, and Jed?




4. Jed plans to marry Elly May, Ned's sister. The risk that a daughter of theirs would inherit nephrolithiasis is _____ and the risk that a son of theirs would inherit the condition is _____ .

# PRENATAL MICROARRAY SCREEN

## KEY WORDS

DNA microarray
Mendelian inheritance
Mode of inheritance

Penetrance
Population genetics

Janelle and Spencer are in their mid twenties, and are planning to have children soon, so they take a panel of tests to identify the most common alleles of the most common autosomal recessive conditions. For each disease, a microarray includes a DNA probe representing the wild type allele bound to a green fluorescent label, and the most common disease-causing allele for the African American population bound to a red fluorescent label. Then a DNA sample is applied to the probes on the microarray. Lighting up both signals for a particular gene indicates that the person is a heterozygote (carrier); lighting up only the red probe indicates that the person is homozygous recessive for the disease-causing allele.

These are the results for Janelle and Spencer:

| Disease | ● = mutant allele | | 0 = normal allele | |
|---|---|---|---|---|
| cystic fibrosis | green 0 | green 0 | green 0 | green 0 |
| beta thalassemia | red ● | green 0 | red ● | green 0 |
| hemochromatosis | green 0 | green 0 | red ● | green 0 |
| osteogenesis imperfecta | green 0 | green 0 | green 0 | green 0 |
| adenosine deaminase deficiency | red ● | green 0 | green 0 | green 0 |
| albinism | red ● | green 0 | red ● | green 0 |
| | Janelle | | Spencer | |

(Cystic fibrosis causes thickened secretions that impair breathing and digestion. Beta thalassemia is a form of anemia. Hemochromatosis causes iron overload that damages several organs. Osteogenesis imperfecta causes brittle easily broken bones. Adenosine deaminase deficiency is an immune deficiency. Albinism is a lack of pigmentation.

## WORKSHEET:

1. Why is it important to know the penetrance of the illnesses?

2. Janelle is an African-American, and Spencer is Caucasian and of English heritage. Why might the particular disease-causing alleles in their microarrays be different?

3. What information about the interactions of alleles is important to know to interpret the results of this test?

4. A DNA microarray test for an autosomal dominant condition with complete penetrance is unnecessary because _____

_____.

5. Which disorders does a child of Janelle and Spencer face a 1 in 4 chance of developing?

6. The chance that their child will be a carrier for hemochromatosis is _____.

7. The chance that a child of theirs will be a carrier of cystic fibrosis and a carrier of adenosine deaminase deficiency is _____.

8. Explain how false negatives can result from the tests that these two people undergo.

# PSEUDOHERMAPHRODITISM

## KEY WORDS

Consanguinity
Genetic code
Genetic drift

Genotype
Mutation
Phenotype

In 29 families in the Dominican Republic, a single base change in a gene that encodes the enzyme 5-alphareductase renders genital tissues in males unable to respond to the male sex hormone testosterone. The mutant allele is recessive and located on chromosome 2. Homozygous recessive females have a normal phenotype.

Homozygous recessive males have a normal male chromosome constitution of XY, but they have ambiguous genitalia, and may look more like a girl than a boy. Their testicles do not descend, and they may have a vagina-like pouch near a very small penis. They are locally called "guevodoces," which means "penis at twelve" in Spanish. At puberty, they develop some male secondary sexual characteristics, such as a deep voice, heavy musculature, and a larger penis. Before the condition was understood, affected individuals were raised as girls until puberty, and then as boys.

All 29 families have the same mutation, a change in the mRNA that encodes the 246$^{th}$ amino acid from CGG to UGG. This society is geographically and socially isolated, and relatives often marry.

**SOURCE:** OMIM 264300

**WORKSHEET:**

1. Why would it be difficult to construct a pedigree for this condition?


2. The amino acid change that causes pseudohermaphroditism is _____ to _____.


3. The DNA sequence of the region affected by the mutation is changed from _____ to _____.

117

4. A reason why this disorder is prevalent in this population is that

_____.

5. The fact that all 29 families have the same mutation suggests that

_____.

6. The fact that the phenotype is not expressed in females illustrates:

    a. sex-influenced inheritance.
    b. sex-linked inheritance.
    c. sex-limited inheritance.
    d. genomic imprinting.
    e. sexual harassment.

7. Why would testosterone shots not help young affected boys develop normally?



8. Is this form of pseudohermaphroditism incompletely penetrant? How do you know this?

# SILVER-RUSSELL SYNDROME

## KEY WORDS

Chromosome aberrations
Genomic imprinting

Microsatellites
Uniparental disomy

Russell-Silver syndrome causes poor growth before and after birth. About 7% of cases are caused by uniparental disomy of a gene or genes on chromosome 7. Marie Alvarez was born with the condition because she inherited her two copies of chromosome 7 from her mother. Christopher Clark, however, is also affected, but has a duplication in a small section of chromosome 7p that microsatellite marker analysis showed was of maternal origin, plus a deletion for this region on his chromosome 7 inherited from his father.

**SOURCES:** OMIM 180860

Monk, David. January 2000. Duplication of 7p11.2-p13, including GRB10, in Silver-Russell syndrome. *The American Journal of Human Genetics* 66:36.

## WORKSHEET:

1. How can Marie and Christopher have the same syndrome but different causative chromosomal aberrations?

2. What is the difference between a duplication and uniparental disomy, in terms of the location of the genes on the chromosomes?

3. Two types of chromosome aberrations that could cause a duplication are

_____ and _____.

4. What types of reproductive problems might Marie and Christopher encounter should they decide to have children together?

# SMITH-LEMLI-OPITZ SYNDROME (RSH SYNDROME)

**KEY WORDS**

Balanced polymorphism
Inborn error of metabolism
Mode of inheritance

Pedigree
Pleiotropy

Zachary was born with a very small head and unusual facial features that suggested a genetic disorder. He had low-set ears, a triangular head, and a small upturned nose. His genitals were so small that nurses in the delivery room at first thought he was a she. Once his parents took him home, they soon noticed that Zachary had difficulty eating, and often emitted a high-pitched scream. He did not gain very much weight. His fingers were slightly webbed, and by the time that he was ready to enter preschool, his parents suspected that he might be mentally retarded.

Zachary's parents and older sister are healthy, but his paternal uncle is mentally retarded and has webbing between his toes. Eventually, Zachary was diagnosed with an autosomal recessive inborn error of metabolism, Smith-Lemli-Opitz syndrome. He lacks an enzyme that is part of the biochemical pathway to synthesize cholesterol. The cholesterol precursor 7-dehydrocholesterol builds up in his bloodstream and presumably caused his symptoms, but he has very low serum cholesterol.

**SOURCE:** OMIM 270400

**WORKSHEET:**

1. Draw a pedigree for Zachary's family, indicating known carriers of Smith-Lemli-Opitz syndrome.

2. What is the evidence that indicates that Zachary's condition probably did not arise as a new mutation?

3. Devise an experiment to test the hypothesis that carriers of this disorder have a lower than normal risk of developing cardiovascular disease.

4. Smith-Lemli-Opitz syndrome is a genetic disease, but may protect against another illness. The general term for this phenomenon is

_____.

5. The risk that Zachary's sister Beth is a carrier is _____.

6. Does the information provided indicate that this condition is incompletely penetrant, variably expressive, or both?

7. What is the evidence that Smith-Lemli-Opitz syndrome is pleiotropic?

# STEM CELLS IN THE HEART

## KEY WORDS

Differentiation
FISH
Stem cells

Transplant
Y chromosome

Sex mismatched organ transplants present an opportunity for researchers to watch how donor and recipient cells interact, because only one of the pair has cells with easily-detectable Y chromosomes. This was the case for a study of eight men who had died months to years after receiving donor hearts from women, a procedure that placed the smaller female hearts under great stress.

The researchers used fluorescence *in situ* hybridization (FISH) to highlight the Y chromosomes in cells in or near the new organ. They sampled cells from all over the donor hearts, as well as from the remnant of heart tissue left in the recipient to attach to the incoming organ. Specifically, they looked 16,834 cardiac muscle cells (myocytes), 25,642 smooth muscle cells that form coronary arterioles (small blood vessels), and 15,539 endothelial cells from capillaries, and checked them for:

- Y chromosomes (indicating recipient cells)
- three cell surface antigens that together indicate a stem cell
- cell surface antigens that distinguish myocyte from smooth muscle cell from endothelial cell
- proteins that indicate cell division

They also studied cells from six male and four female healthy hearts to see whether stem cells are normally in the organ.

The researchers discovered cells in the recipient's new heart that had Y chromosomes and the large nuclei and cell surface characteristics of stem cells. They estimated that recipient cells account for about 18 percent of the myocytes, 20 percent of the coronary arteriole smooth muscle cells, and 14 percent of the endothelium of coronary capillaries in the new hearts. The male cells appeared in all eight hearts in similar spatial patterns. Normal hearts had about a quarter of the stem cells as the transplanted hearts

The experiments also showed that infiltration of recipient cells into the donor heart occurs rapidly. Y-bearing cells were found in the heart of one recipient who had lived only four days after surgery!

**SOURCE:** Quaini, Federico et al. January 3, 2002. Chimerism of the transplanted heart. *The New England Journal of Medicine* 346(1):5.

**WORKSHEET:**

1. Not many human tissues are known to regenerate. Physicians in training have been taught that the liver can regenerate, but the heart cannot. What is the evidence that this teaching will have to change?

2. The same research group that conducted the study on the eight transplant recipients had a year earlier discovered that certain bone marrow stem cells can repopulate a heart damaged by a heart attack, differentiating into the variety of cell types required to knit new vessels and valves, muscle and connective tissue. Devise an experiment to distinguish between two hypotheses: (1) that new heart tissue in a transplant recipient comes from bone marrow cells or (2) that new heart tissue comes from stem cells in the remaining tissue from the recipient's own heart.

3. Many people object to using stem cells from embryos or fetuses to treat medical conditions in adults. How does this study on heart transplants possibly render their objections moot?

4. The study could not trace the response of a female recipient to a transplant from another female because:




5. Why might there be more stem cells in the female hearts transplanted into male chests, than in normal female hearts?




6. The eight patients survived for different times. Those who lived the longest had the fewest stem cells, yet had the most Y-bearing cells that showed signs of both division and differentiation. Describe the overall process that this evidence suggests seems to be unfolding as the recipient's body adapts to the new organ.




7. A recipient's stem cells and differentiated cells in a donor organ is a good sign. What is the alternative?




8. What is a clinical implication or application of the results of these experiments?

# TANGIER DISEASE

## KEY WORDS

Gene therapy
Genotype
Mendel's second law
Phenotype
Population genetics

Most of the 50 known people who have Tangier disease live on Tangier island in the Chesapeake Bay – hence the name. Most are descendants of the original settlers who arrived in 1686. The disease is also called analphalipoproteinemia after its biochemical characteristic – low blood serum levels of high density lipoprotein (HDL), which normally transports cholesterol from the blood to the liver, where it is metabolized. HDL is the "good" cholesterol. Tangier disease is inherited as an autosomal recessive trait. People with the condition have low blood serum cholesterol, but excess cholesterol in other areas, such as the thymus gland and in scavenger cells of the immune system called macrophages. Other, unexplained symptoms include:

- large, orange tonsils
- inability to sense pain, heat, and cold on the skin
- wasting of muscles in the hands
- enlarged liver, spleen, and lymph nodes
- premature coronary artery disease

In 1998, researchers traced Tangier disease to a gene on the long arm of chromosome 9 that encodes a protein called an ATP binding cassette transporter (*ABCA1*). This is part of the protein complex that brings in ATP to power HDL's movement of cholesterol. A one base deletion in the gene creates a premature stop codon 24 bases away. The result is an extremely shortened protein product. It lacks the portions that bind to ATP and to liver cell membranes.

Ronald is a resident of Tangier island who has the disease but feels well. He marries Nancy, a neighbor whose mother has Tangier disease but whose father does not. Nancy does not have Tangier disease, but she has inherited the heterozygous form of familial hypercholesterolemia (FH) from her father, who died at a young age of a heart attack. In FH, an autosomal dominant condition, liver cells have half the normal number of low density lipoprotein (LDL) receptors, and cholesterol accumulates in the blood, causing early heart disease. Nancy knows that she has probably inherited heterozygous FH because her blood serum cholesterol is dangerously high without medication and strict dietary control.

**SOURCES:** OMIM 205400, 600046

Bodzioch, M. et al. August 1999. The gene encoding ATP-binding cassette transporter 1 is mutated in Tangier disease. *Nature Genetics* 22:347.

Brooks-Wilson, A. et al. August 1999. Mutations in *ABC1* in Tangier disease and familial high density lipoprotein deficiency. *Nature Genetics* 22:336.

Rust, S. et al. August 1999. Tangier disease is caused by mutations in the gene encoding ATP-binding cassette transporter 1. *Nature Genetics* 22:352.

**WORKSHEET:**

1. The mutation that causes Tangier disease is:
    a. missense
    b. nonsense
    c. antisense
    d. trisomy
    e. silent

2. A healthy heart profile is to have high blood serum HDL and low LDL. Would inheriting both of these conditions cancel each other out, or make a person about twice as likely to suffer from a complication of high blood serum cholesterol?

3. Assuming that Tangier disease and FH are transmitted on different chromosomes, the chance that a child of Nancy and Ronald inherits both disorders is _____.

4. Even though we know the gene and gene product for Tangier disease, gene therapy would be difficult to develop because _____

_____.

5. A genetic phenomenon that explains why Tangier disease is very common on this island but is rare elsewhere is _____.

# TOWNES-BROCKS SYNDROME

## KEY WORDS

Autosomal dominant
Autosomal recessive
Mode of inheritance
Pedigree
Pleiotropy
X-linked recessive

Individuals who inherit Townes-Brocks syndrome have:

- tags of extra skin on the ears and hearing loss
- absent or extra skin on the anus
- a doubled thumb and webbing between fingers
- overlapping toes

The Dominguez family depicted in the pedigree below has several members who are affected to different degrees, as is indicated by different shaded areas on the pedigree symbols:

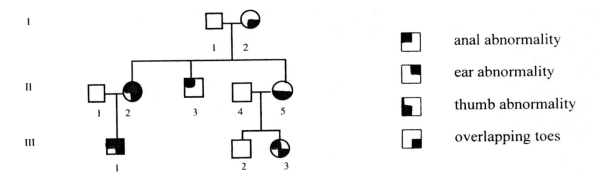

**SOURCE:** OMIM 107480

## WORKSHEET:

1. If just overlapping toes is considered as a separate trait, which modes of inheritance are possible in this family?

2. Which mode of inheritance applies if all symptoms are considered as part of the same syndrome?




3. If individual II3 has children with a woman who does not have the syndrome, how does the chance that a son inherits the condition differ if it is inherited as an autosomal recessive, autosomal dominant, or X-linked recessive trait?




4. Several other families with this condition also exhibit male-to-male transmission. If these families are considered along with the Dominguez's, what is the most likely mode of inheritance?

# VENOUS THROMBOSIS

## KEY WORDS

Expressivity
Genetic heterogeneity
Penetrance

Relative risk
Translation

Many genes and their encoded protein factors are part of the biochemical pathway that enables blood to clot. Below is a simplified version of part of this complex pathway:

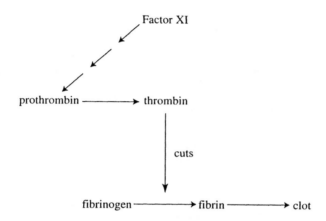

For blood to clot efficiently, factor XI activates a series of other factors, which convert prothrombin to thrombin. Thrombin then cuts a protein called fibrinogen to the shorter fibrin, which assembles into the threads of a clot. Blood clotting is a normal process to heal an injury, but may be activated abnormally if the linings of blood vessels are not smooth, which can happen if a person eats a fatty diet for many years and does not exercise.

In factor XI deficiency, a person bleeds easily upon injury. Symptoms vary from just taking slightly longer to stop bleeding, to serious blood loss. In some people, symptoms may be so mild that the condition may not be noticed unless major injuries occur.

Researchers reasoned that if too little factor XI impairs clotting, then perhaps excess or too-active factor XI might contribute to illness caused by too-efficient clotting, such as venous thrombosis. In this condition, clots form in the veins, blocking circulation. The researchers examined 434 individuals who had venous thrombosis, and compared them to 434 individuals matched for sex and age who did not have the condition. The study revealed that individuals above the 90$^{th}$ percentile for factor XI level in the blood had a relative risk of 2.2% of developing venous thrombosis. The protocol ruled out contributions from other risk factors, including oral contraceptive use, poor diet, and mutations in genes that encode other clotting factors, leukocyte adhesion molecules, or enzymes that regulate homocysteine metabolism.

**SOURCE**: Meijer's, Joost C. M. et al. March 9, 2000. High levels of coagulation factor XI as a risk factor for venous thrombosis. *The New England Journal of Medicine* 342:69.

**WORKSHEET:**

1. How can mutations in the same gene lead to opposite phenotypes – that is, too much or too little blood clotting?

2. Which risk factors for cardiovascular disease are controllable and which are not?

3. Which of the following statements are true?

    a. Factor XI deficiency is variably expressive and completely penetrant.
    b. Factor XI deficiency is variably expressive and incompletely penetrant.
    c. Poor blood clotting is a genetically heterogeneic trait.
    d. Factor XI deficiency is X-linked recessive.
    e. A person with factor XI deficiency has twice the risk of developing venous thrombosis as a person with normal factor XI activity.

4. A mutation in the gene that encodes thrombin that decreases its activity would cause a phenotype of _____.

5. A participant in the study reads the published report, and determines that because she is in the 96th percentile for factor XI activity, that her risk of developing venous thrombosis is only 2.2%. How is she incorrect?

6. Is the formation of mature fibrin an event that occurs before or after translation of mRNA into protein?




7. How does a person's environment contribute to the manifestation of factor XI deficiency?

# APPENDIX A

## The Genetic Code

| First Letter | Second Letter | | | | Third Letter |
|---|---|---|---|---|---|
| | U | C | A | G | |
| U | UUU } phenylalanine (phe)<br>UUC }<br>UUA } leucine (leu)<br>UUG } | UCU }<br>UCC } serine (ser)<br>UCA }<br>UCG } | UAU } tyrosine (tyr)<br>UAC }<br>UAA STOP<br>UAG STOP | UGU } cysteine (cys)<br>UGC }<br>UGA STOP<br>UGG tryptophan (try) | U<br>C<br>A<br>G |
| C | CUU }<br>CUC } leucine (leu)<br>CUA }<br>CUG } | CCU }<br>CCC } proline (pro)<br>CCA }<br>CCG } | CAU } histidine (his)<br>CAC }<br>CAA } glutamine (gln)<br>CAG } | CGU }<br>CGC } arginine (arg)<br>CGA }<br>CGG } | U<br>C<br>A<br>G |
| A | AUU }<br>AUC } isoleucine (ilu)<br>AUA }<br>AUG+ methionine (met) | ACU }<br>ACC } threonine (thr)<br>ACA }<br>ACG } | AAU } asparagine (asn)<br>AAC }<br>AAA } lysine (lys)<br>AAG } | AGU } serine (ser)<br>AGC }<br>AGA } arginine (arg)<br>AGG } | U<br>C<br>A<br>G |
| G | GUU }<br>GUC } valine (val)<br>GUA }<br>GUG } | GCU }<br>GCC } alanine (ala)<br>GCA }<br>GCG } | GAU } aspartic acid (asp)<br>GAC }<br>GAA } glutamic acid (glu)<br>GAG } | GGU }<br>GGC } glycine (gly)<br>GGA }<br>GGG } | U<br>C<br>A<br>G |

# APPENDIX B

**Symbols:**

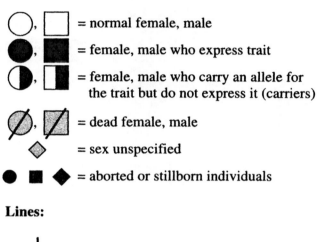

○, □ = normal female, male
●, ■ = female, male who express trait
◐, ◨ = female, male who carry an allele for the trait but do not express it (carriers)
⌀, ⌀ = dead female, male
◇ = sex unspecified
● ■ ◆ = aborted or stillborn individuals

**Lines:**

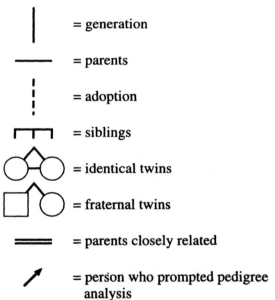

| = generation
— = parents
⋮ = adoption
⊓ = siblings
= identical twins
= fraternal twins
= parents closely related
↗ = person who prompted pedigree analysis

**Numbers:**

Roman numerals = generations

Arabic numerals = individuals

Symbols used in pedigree construction are connected to form a pedigree chart, which displays the inheritance patterns of particular traits.